A MANUAL OF PROBLEMS IN STRUCTURAL GEOLOGY

A MANUAL OF PROBLEMS IN STRUCTURAL GEOLOGY

By

N.W. Gokhle

M.Sc., Ph.D. (Budapest)

F.M.S. (Mysore), F.I.S.E.S. (Pune)

Professor of Geology,

Department of Studies in Geology, Karnataka University Dharwad-58003

CBS

CBS Publishers & Distributors Pvt. Ltd.

New Delhi • Bengaluru • Chennai • Kochi • Kolkata • Mumbai

Hyderabad • Uttarakhand • Nagpur • Patna • Pune • Jharkhand

ISBN: 81-239-1303-6

First Edition: 1991
Reprint: 2000, 2006, 2008, 2009, 2011, 2012, 2014, 2016, 2018

Published by **Satish Kumar Jain** and produced by **Varun Jain** for
CBS Publishers & Distributors Pvt. Ltd.,
4819/XI Prahlad Street, 24 Ansari Road, Daryaganj, New Delhi - 110002
delhi@cbspd.com, cbspubs@airtelmail.in • www.cbspd.com
Ph.: 23289259, 23266861, 23266867 • Fax: 011-23243014

Corporate Office: 204 FIE, Industrial Area, Patparganj, Delhi - 110 092
Ph: 49344934 • Fax: 011-49344935
E-mail: publishing@cbspd.com • publicity@cbspd.com

Branches:
- ***Bengaluru:*** 2975, 17th Cross, K.R. Road, Bansankari 2nd Stage, Bengaluru - 70 • Ph: +91-80-26771678/79 • Fax: +91-80-26771680 E-mail: cbsbng@gmail.com, bangalore@cbspd.com
- ***Chennai:*** No. 7, Subbaraya Street, Shenoy Nagar, Chennai - 600030 Ph: +91-44-26681266, 26680620 • Fax: +91-44-42032115 E-mail: chennai@cbspd.com
- ***Kochi:*** Ashana House, 39/1904, A.M. Thomas Road, Valanjambalam, Ernakulum, Kochi • Ph: +91-484-4059061-65 Fax: +91-484-4059065 • E-mail: cochin@cbspd.com
- ***Kolkata:*** 6-B, Ground Floor, Rameshwar Shaw Road, Kolkata - 700014 Ph: +91-33-22891126/7/8 • E-mail: kolkata@cbspd.com
- ***Mumbai:*** 83-C, Dr. E. Moses Road, Worli, Mumbai - 400018 Ph: +91-9833017933, 022-24902340/41 • E-mail: mumbai@cbspd.com

Representatives:
- Hyderabad: 0-9885175004
- Patna: 0-9334159340
- Jharkhand: 0-9811541605
- Nagpur: 0-9021734563
- Pune: 0-9623451994
- Uttarakhand: 0-9716462459

Printed at:
Neekunj Print Process, Delhi (India)

PREFACE

The book describes several types of problems arising in the field of practical structural geology. These comprise in the main of ascertaining the exact attitude of the geologic body from the data collected in the field. This requires the determination of true dip amount and direction, if the geologic body is bedded one. For the estimation of ore reserves, the size and shape are required to be determined. For this purpose the depth and thickness of the geologic body is needed to be determined. Bearing this in mind several methods have been described in the book, of determining true dip and apparent dip from several modes of setting the data.

The geologic bodies are often deformed by folding and faulting. It becomes necessary therefore of ascertaining the structure as to whether it is a syncline, an anticline, plunging fold and so on. The amount of throw and down throw side also need to be determined. Borehole data are essential in such problems. The book elaborates the procedure of handling such data.

Stereographic projections are very useful in handling inclination of stratified bodies given in degrees. Solution can be quickly obtained by adopting the procedure of sterographic projection and this has been described in the book.

To a geoscientist, clinometer compass is an indispensible instrument. The operation part of it and especially why the east and west directions are reversed in a clinometer compass have been elaborated. Guidelines have been set out about the usage of this instrument.

The book will be found useful because several aspects of practical structural geology have been described together in one volume, the need for which is felt by all those who desire to learn the methodology. The information is presented in a systematic manner, covering the minutest details considering hypothetical problems. The text though not written according to the syllabus of any university, the requirements of students are not lost sight of.

DHARWAD

N.W. GOKHALE.

CONTENTS

1

CALCULATION OF TRUE AND APPARENT DIP

1.1.0 INTRODUCTION

The geoscientists are often called upon to establish the general true dip from the availability of apparent dips as recorded at several places in the field. In the laboratory atleast, the dip is expressed in the form of a ratio; like 1 : 5, 1 : 10 etc. Also the depth to a certain horizon as obtained by drilling at several spots, and from such a data the geoscientist is called upon to calculate the true dip. On other occasions, the depth to a geologic body is required to be estimated in a specific direction of traverse. Obviously it becomes necessary to calculate the amount of dip (may be apparent one) in the given direction of traverse. To establish the existence of a fold (anticline or syncline), the given data of apparent dips, needs to be analysed, and the direction of true dip derived.. In the case of faults, the amount of throw and the side of downthrow can be established only after calculating true dip, which inturn helps to calculate the expected level of the occurrence of a particular horizon, across the fault. In all such problems, the starting point is true dip, its direction and the amount. In the following pages, the procedure of calculating true dip, apparent dip, etc., will be detailed out.

1.1.1 TRUE DIP AND APPARENT DIP

Dip or inclination is possessed by the sedimentary beds, tabular bodies like dykes, sills, lava flows, and so on. True dip is the maximum inclination of a dipping tabular body. Apparent dip is one which is measured in a direction other than that of the maximum dip. Value of apparent dip varies between the maximum value (true dip) and zero degrees. The direction along which *there is no (zero) inclination* is called as the "true strike" direction of the dipping body. In fact in order to ascertain this, the direction of *zero or no inclination* should be located first. Perpendicular to such a direction, the dip will be found to be maximum.

1.1.2 DETERMINATION OF TRUE DIP FROM APPARENT DIP

There are several ways of setting the data of apparent dip. These are described below.

1.1.3 DIP GIVEN AS A GRADIENT

Problem 1.1.0 The apparent dips were recorded in a sandstone quarry namely 1 : 5 due S 20 °E and 1 : 11 due N 60 °E. Find the direction and amount of true dip.

Procedure

From point *O*, draw lines *OA* and *OB* due S 20 °E and N 60 °E respectively (Fig. 1.1.1.). Choose a convenient unit distance to accomodate apparent dip amounts of 5 and 11. Bigger the unit, more accurate will be the result. Mark off 5 and 11 units along the lines *OA* and *OB* respectively and obtain points *A′* and *B′*. Since points *A′* and *B′* are one unit lower than point *O*, the points *A′* and *B′* are therefore having equal elevation. A line joining the points *A′* and *B′* therefore describes a horizontal plane (in the same manner as a contour line which connects the points of equal elevation, and therefore it is contained in a horizontal plane). Join points *A′* and *B′* by a straight line and extend it on either sides. This line yields the true strike line. Draw a perpendicular from *O* on this line *A′* B′ and extend it, which has been labelled

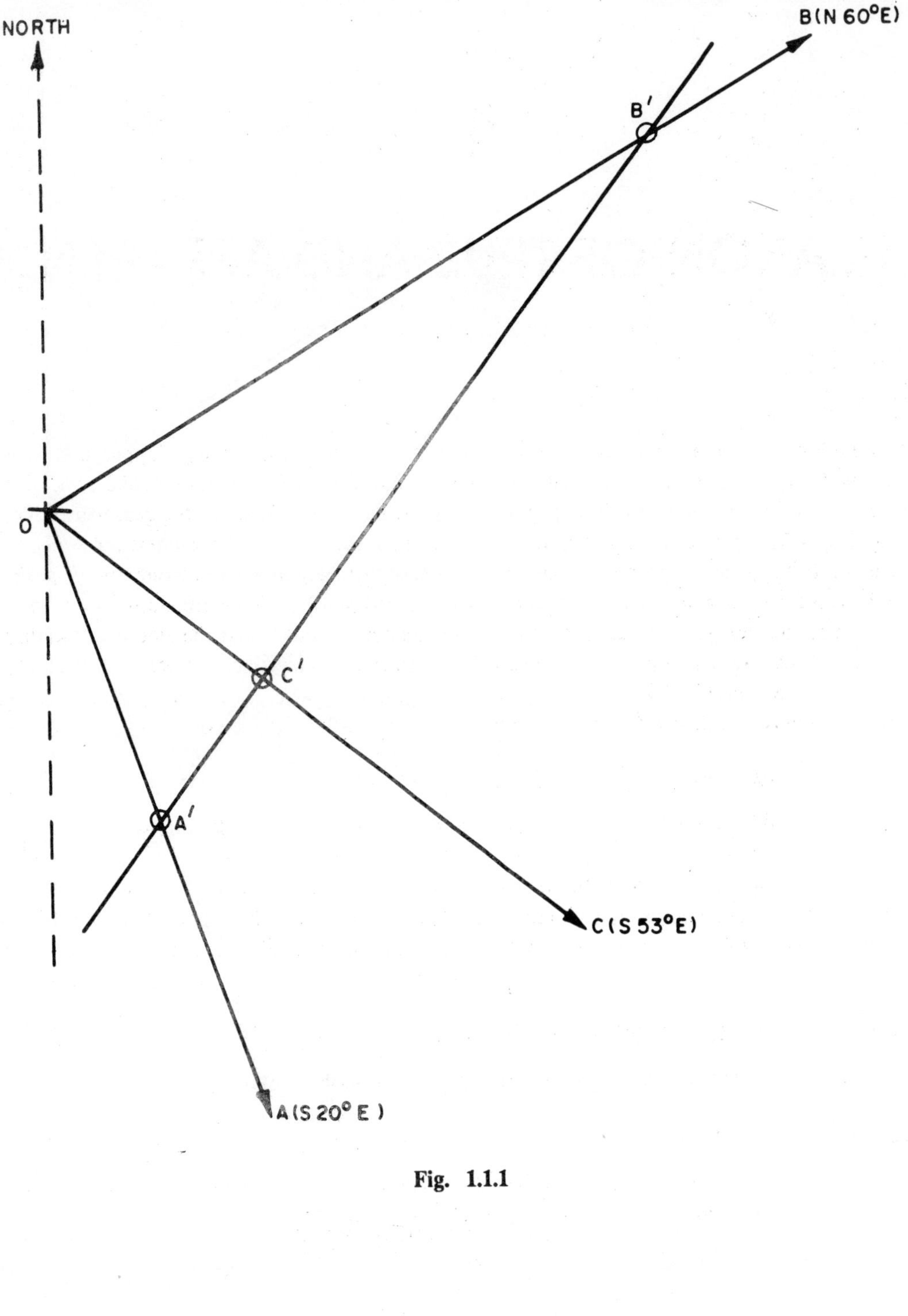

Fig. 1.1.1

Fig. 1.1.1 TRUE DIP FROM TWO APPARENT DIPS (Problem 1.1.0)

OA	=	apparent dip direction, S 20 °E
OA'	=	amount of apparent dip (5 units) along *OA*
OB	=	apparent dip direction, N 60 °E
OB'	=	amount of apparent dip (11 units) along *OB*
A'B'	=	true strike line
OC	=	true dip direction, S 43 °E
OC'	=	true dip amount (4.2 units) along *OC*
		Scale 1 cm. = 1 unit

as line *OC in Fig. 1.1.1*. *OC* is the direction of true dip and the distance *OC′* measured to the same unit length as taken for the distance OA′ and *OB′*, gives the amount of true dip. It is found that the direction of the true dip (*OC*) is S 53 °E and the amount of true dip is 1 :4.2 (distance *OC′*).

1.1.4 GRADIENT GIVEN IN MODIFIED FORM

It is not necessary to give two directions of apparent dips. It is yet possible to calculate the true dip direction and amount from one direction of apparent dip and other data. The variations in setting the data are given below.

Problem 1.1.1 A bedded formation possesses an apparent dip of 1 : 10 due N 25 °W. The dirction of true strike is N 40 °W–S 40 °E. Determine the direction and amount of true dip.

Procedure

From point *O*, draw a line *OA* due N 25 °W and mark off 10 units on it from *O*, and obtain point *A′*. Through *A′* draw a line trending N 40 °W–S 40 °E, which is the given direction of true strike. Drop a perpendicular *OB* from point *O* on the true strike line. This line *OB* yields the direction of true dip, which will be found to be due N 50 °E. It is to be noted that if only the dirction of true strike is given, then the dip direction cannot be immediately ascertained because, it could be in one direction, or exactly in the opposite direction. It has therefore two solutions. Thus in the present problem, the dip direction could be due N 50 °E or S 50°W, but it could not be both at one and the same time, unless the formation is folded one. In the case cited in problem 1.1.1, the dip is due N 50 °E. The line *OB* cuts the strike line at *B′*. The distance OB′ measured to the same unit length as taken along *OA′*, gives the amount of true dip. It is found to be 1 : 4.2 (Fig. 1.1.2)

Problem 1.1.2 A formation has an apparent dip of 1 : 10 due S 60 °W. The amount of true dip is 1 : 4. Find the direction of true dip.

Procedure

From point *O* draw a line *OA* 10 units long due S 60 °W, this being the given direction of apparent dip. (Fig. 1.1.3). With *O* as centre draw a circle with a radius of 4 units. From point *A*, draw tangent to the circle inscribed, around *O*. It will be found that two tangents can be drawn, namely *AB* and *AC* which touch the circle at points *B′* and *C′*, respectively. Join *OB′* and *OC′* and extend these lines to points *B″* and *C″* respectively. Then *OB′*B″ and *OC′*C″ are the two directions of true dip. These directions are found to be N 55 °W and S 7 °E. Thus two directions of true dip are possible, when only one direction of apparent dip and amount, and only true dip amount are given.

1.1.5 APPARENT DIP FROM TRUE DIP

Problem 1.1.3 A formation has a true dip of 1 : 5 due east. Determine the amount of apparent dip due N 50 °E.

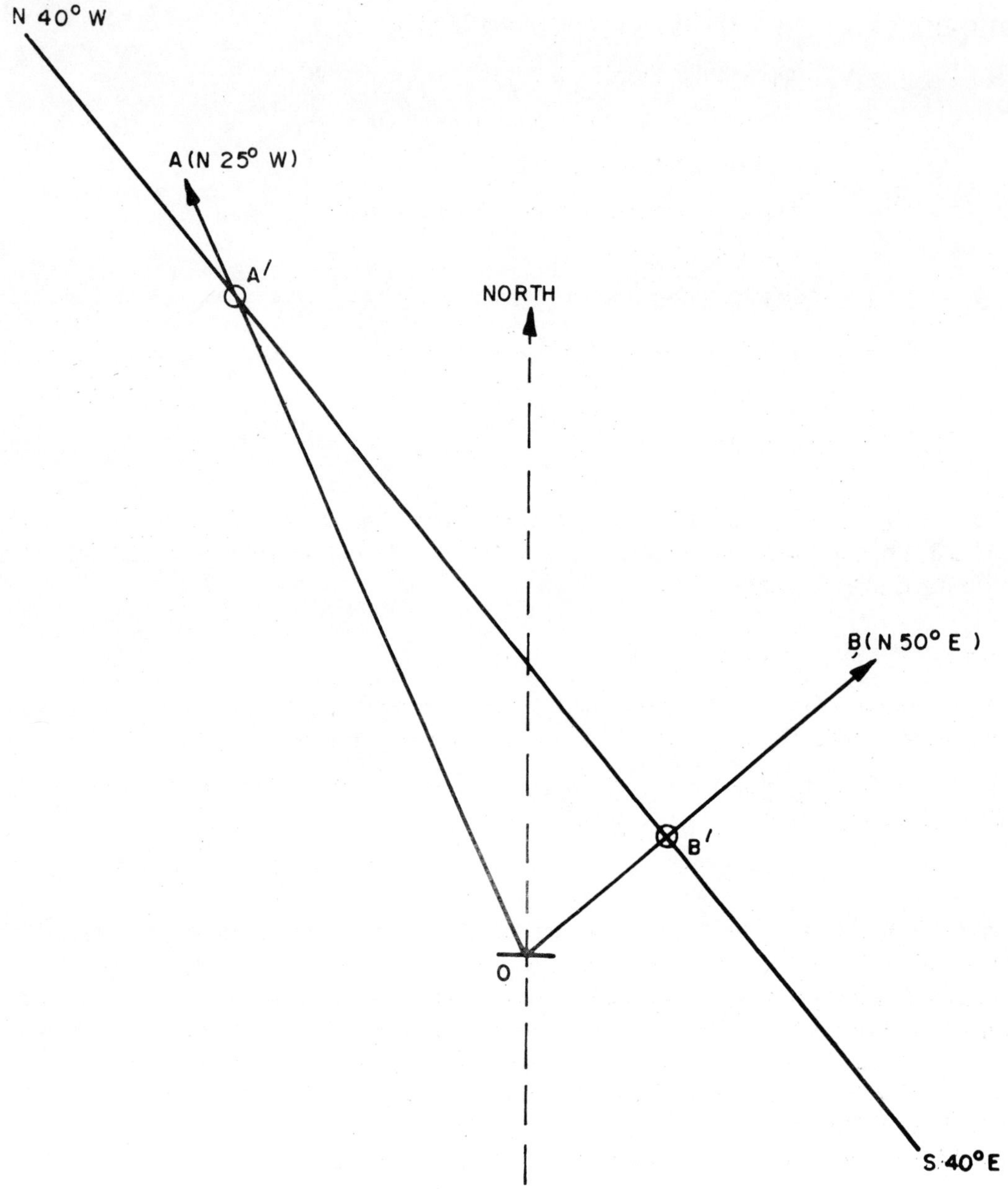

Fig. 1.1.2

Fig. 1.1.2 TRUE DIP FROM ONE APPARENT DIP, AND DIRECTION OF TRUE STRIKE (Problem 1.1.1)

OA	=	apparent dip direction, N 25 °W
OA'	=	apparent dip amount (10 units) along *OA*
A'B'	=	true strike line N 40 °W–S 40 °E
OB	=	true dip direction N 50 °E
OB'	=	true dip amount (2.6 units)

Scale 1 cm = 1 unit

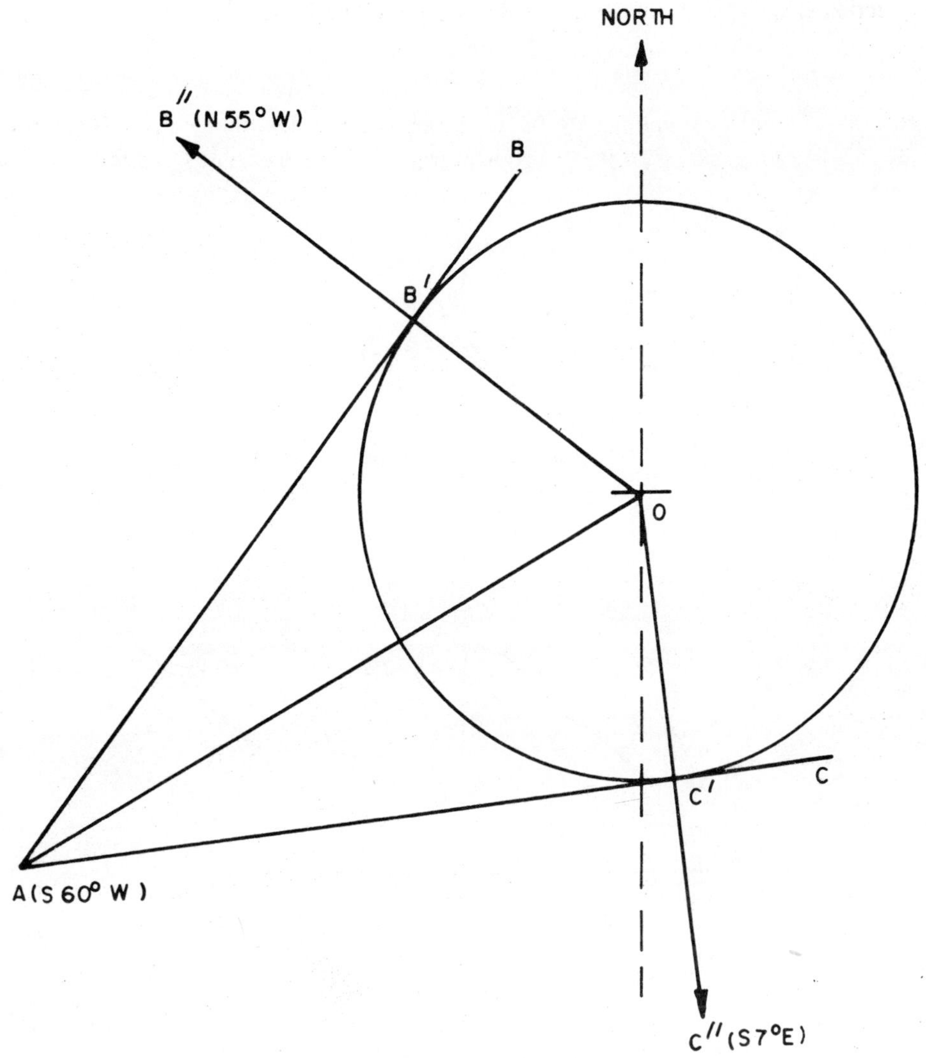

Fig. 1.1.3

Fig. 1.1.3 TRUE DIP DIRECTION FROM TRUE DIP AMOUNT, AND ONE APPARENT DIP (Problem 1.1.2)

OA	=	apparent dip direction, S 60 °W
OA'	=	apparent dip amount (10 units)
AB & AC	=	tangents drawn to the circle having a radius of 4 units, it being the amount of true dip. The tangents touch at the circle at *B* and *C*
OB'B"	=	true dip direction, N 55 °W
OC'C"	=	true dip direction, S 7 °E

Scale 1 cm = 1 unit

Also find out the direction along which the amount of apparent dip will be 1 : 10.

Procedure

From point *O* draw a line *OA* due east, and mark off 5 units along it, and obtain point *A′*. Through *A′* draw a line *XY* perpendicular to *OA*, the line *XY* standing for the direction of true strike. Draw a line *OB* due N 50 °E which is the given

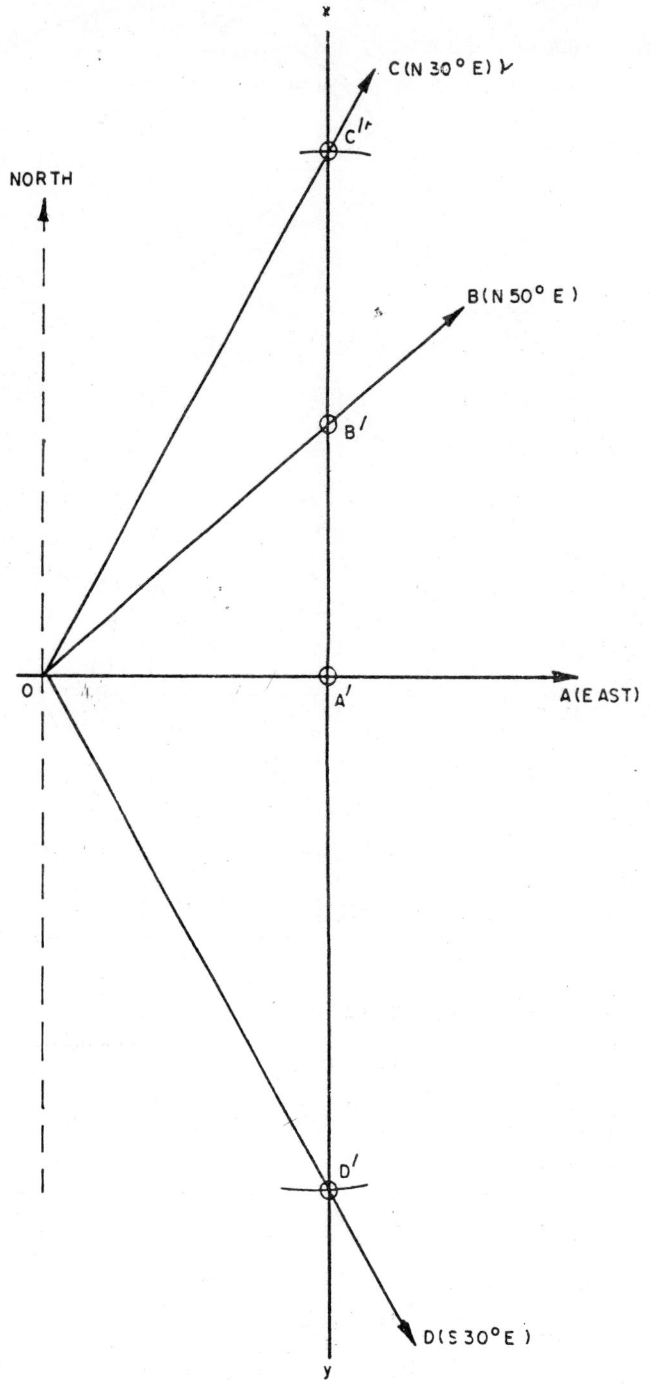

Fig. 1.1.4

Fig. 1.1.4 APPARENT DIP AMOUNT, AND APPARENT DIP DIRECTION, FROM TRUE DIP AMOUNT AND DIRECTION (Problem 1.1.3)

OA	=	true dip direction, east
OA'	=	true dip amount (5 units)
OB	=	apparent dip direction, N 50 °E
OB'	=	apparent dip amount (6.4 units)
OC	=	apparent dip direction, N 30 °E
OC'	=	apparent dip amount (10 units)
OD	=	apparent dip direction, S 30 °E
OD'	=	apparent dip amount (10 units)

Scale 1 cm = 1 unit

direction of apparent dip. This line cuts the true strike line at B'. The length OB' measured according to the unit length chosen, gives the amount of apparent dip due N 50 °E (Fig. 1.1.4).

In order to ascertain the direction of apparent dip along which the amount of apparent dip will be 1 : 10, draw an arc of circle with O as centre and radius of 10 units. This procedure gives two points at which the arc will be intersecting the true strike line (points C' and D' in Fig 1.1.4). Thus two directions are obtained, namely OC' C (N 30 °E) and OD'D (S 30 °E).

1.1.6 DATA OF DIP GIVEN IN DEGREES

The procedure adopted in such cases is different from that described above. Here also the data can be set out in different fashions. The different modes of setting data will be given below.

1.1.7 (a) TRUE DIP FROM TWO APPARENT DIPS

Problem 1.1.4 A formation dips 20° due N 50 °W and 50° due S10 °W. Find the direction and amount of true dip.

Procedure

From point O draw lines OA and OB due N 50 °W and S 10 °W, which are the given directions of apparent dips (Fig. 1.1.5). Construct an angle of 20° with respect to the line OA (angle AOA'), because the amount of apparent dip along OA is 20°. This angle AOA' can be constructed on any side of the line OA. Erect a perpendicular from OA' on OA such that $A''A'''$ is some unit length (1 cm, 1 inch etc.). Like wise construct an angle of 50° at O with respect to line OB (angle BOB') because the amount of apparent dip along OB is 50°. This angle BOB' can be constructed on any side of line OB. Erect a perpendicular from OB' on OB such that $B''B'''$ is equal to the length of the line $A''A'''$. Join the points A''' and B''' by a straight line and extend it on either sides. As the points A''' and B''' are the same one unit lower than the initial point O, the line drawn through point A''' and B''' defines a true strike line. Drop a perpendicular OC on the strike like A''' B''', from point O. This procedure gives the direction of true dip which is found to be S 53 °W. This line OC cuts the strike line at point C'''. From C''' mark off distance C''' - C'' which should be equal to that taken for A'' - A''' or for B'' - B'''. Connect points O and C'' and extend it to C' which gives rise to angle COC'. This latter angle COC' is the amount of true dip along the true dip direction OC. The true dip amount is 59°, the direction of true dip being S 53 °W.

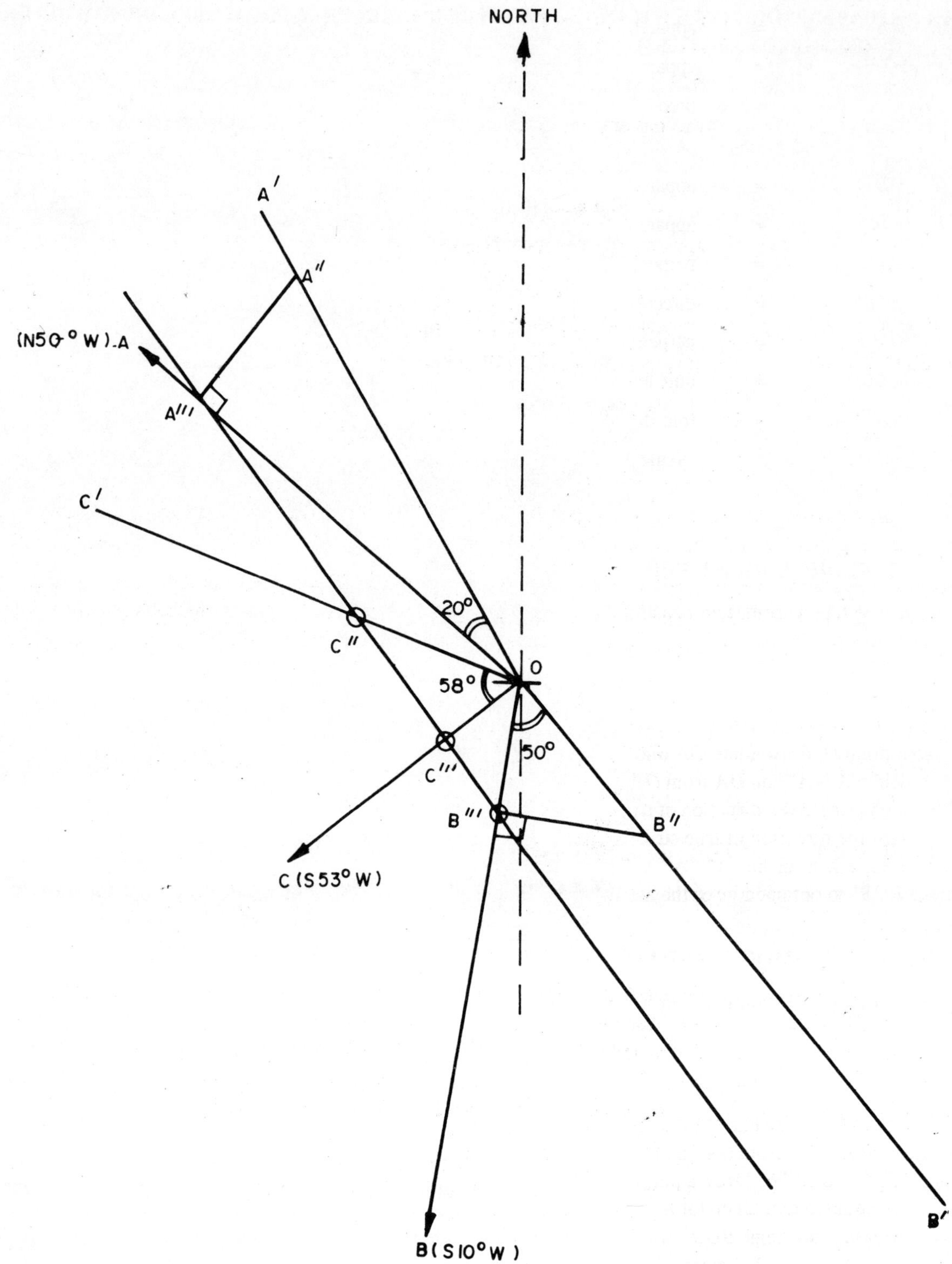

Fig. 1.1.5

Fig. 1.1.5 TRUE DIP FROM TWO APPARENT DIPS (Problem 1.1.4)

OA	=	apparent dip direction (N 50 °W)
AOA'	=	apparent dip amount (20°), along OA
$A''A'''$	=	perpendicular from OA' on OA, length of
		$A''A'''$ = 2.54 cms = 1 unit
OB	=	apparent dip direction (S 10 °W)
BOB'	=	apparent dip amount (50°) along OB
$B''B'''$	=	perpendicular from OB$'$on OB, length of $B''B'''$ = 2.54 cms = 1 unit.
$A'''B'''$	=	direction of true strike = N 37 °W – S 37 °E
OC	=	perpendicular from O on A'''-B''' = true dip direction = S 53 °W.
$C'''C''$	=	unit length = 2.54 cms.
COC'	=	true dip amount along OC = 58°
		Scale = 2.54 cms = 1 unit

1.1.7 (b) TRUE DIP AMOUNT FROM APPARENT DIP AND TRUE DIP DIRECTION

Problem 1.1.5 A formation has an apparent dip of 30° due north. The direction of true dip is N 50 °E. Find the amount of true dip.

Procedure

From point O draw a line OA due north. Construct an angle of 30° with respect to OA at O (angle AOA'). Erect a perpendicular A'' - A''' on OA from OA', choosing some convenient unit length for this line. Draw a line OB due N 50 °E which is the given direction of true dip. From point A''' draw a perpendicular on the line OB and extend it, which line stands for the direction of true strike. This strike line cuts the direction of true dip at point B'''. Lay off a distance of one unit (as taken for line A'' - A$'''$) from point B''', and get point B''. Join points O and B'' and extend this line to B'. The angle BOB' so obtained gives the amount of true dip along OB (Fig 1.1.6). This is found to be 42°.

1.1.7 (c) TRUE DIP AMOUNT AND DIRECTION FROM TRUE STRIKE AND ONE APPARENT DIP

Problem 1.1.6 A formation dips 30° due S 80 °E. The direction of true stike is N 50 ° E - S 50 °W. Find the amount and direction of true dip.

Procedure

From point O draw a line OA due S 80° E. Construct an angle of 30° with respect to line OA at O (angle AOA'). Erect a perpendicular A'' - A''' from OA' on OA, choosing any standard length for this line. Through point A''' draw a line XA''' Y trending N 50 °E - S 50 °W. Draw a perpendicular OB on XA''' Y which cuts the strike line (XA''' Y) at B$'''$. From B$'''$ lay off a distance equal to that taken for A''A$'''$, along the strike line and obtain point B''. Join points O and B'' and extend the line to B'. Measure the angle BOB' which gives the amount of true dip (37°), and the line OB gives the direction (S 40 °E) of true dip. It is be noted that merely from the given direction of true strike of N 50 °E - S 50 °W, the direction of true dip cannot be derived as it has two solutions - either due N 40 °W or S 40 °E. The given direction of apparent dip therefore fixes up the actual direction of true dip. In the present case it is due S 40 °E and not due N 40 °W.

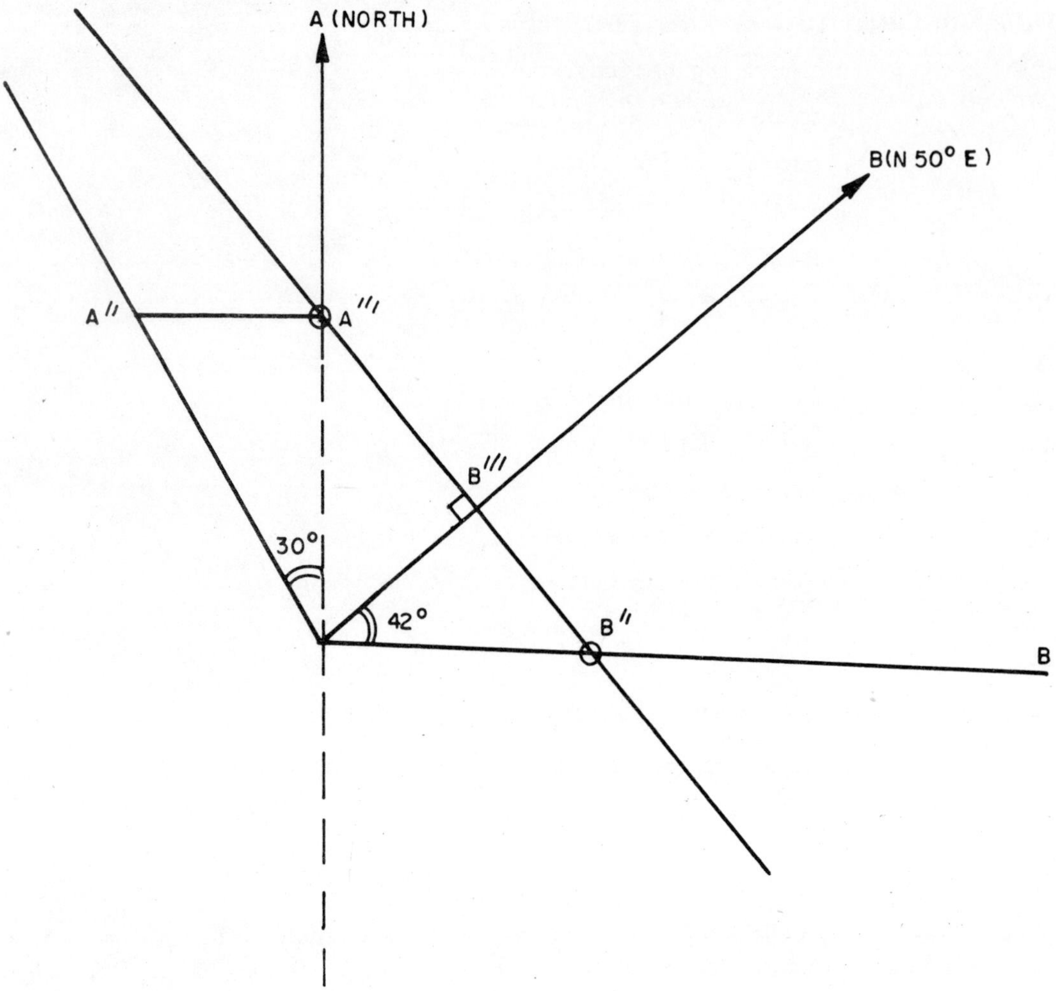

Fig. 1.1.6

Fig. 1.1.6 TRUE DIP AMOUNT FROM ONE APPARENT DIP AND TRUE DIP DIRECTION

OA	=	apparent dip direction, north
AOA'	=	apparent dip amount of 30°, along *OA*
A''A'''	=	perpendicular from *OA'* on *OA* = unit length (2.54 cms)
OB	=	true dip direction, N 50 °E
A'''B'''	=	perpendicular from *A'''* on *OB* = true strike direction.
B'''B''	=	unit length = 2.54 cms.
BOB'	=	true dip amount = 42°, = along *OB*

Scale 2.54 cms = 1 unit

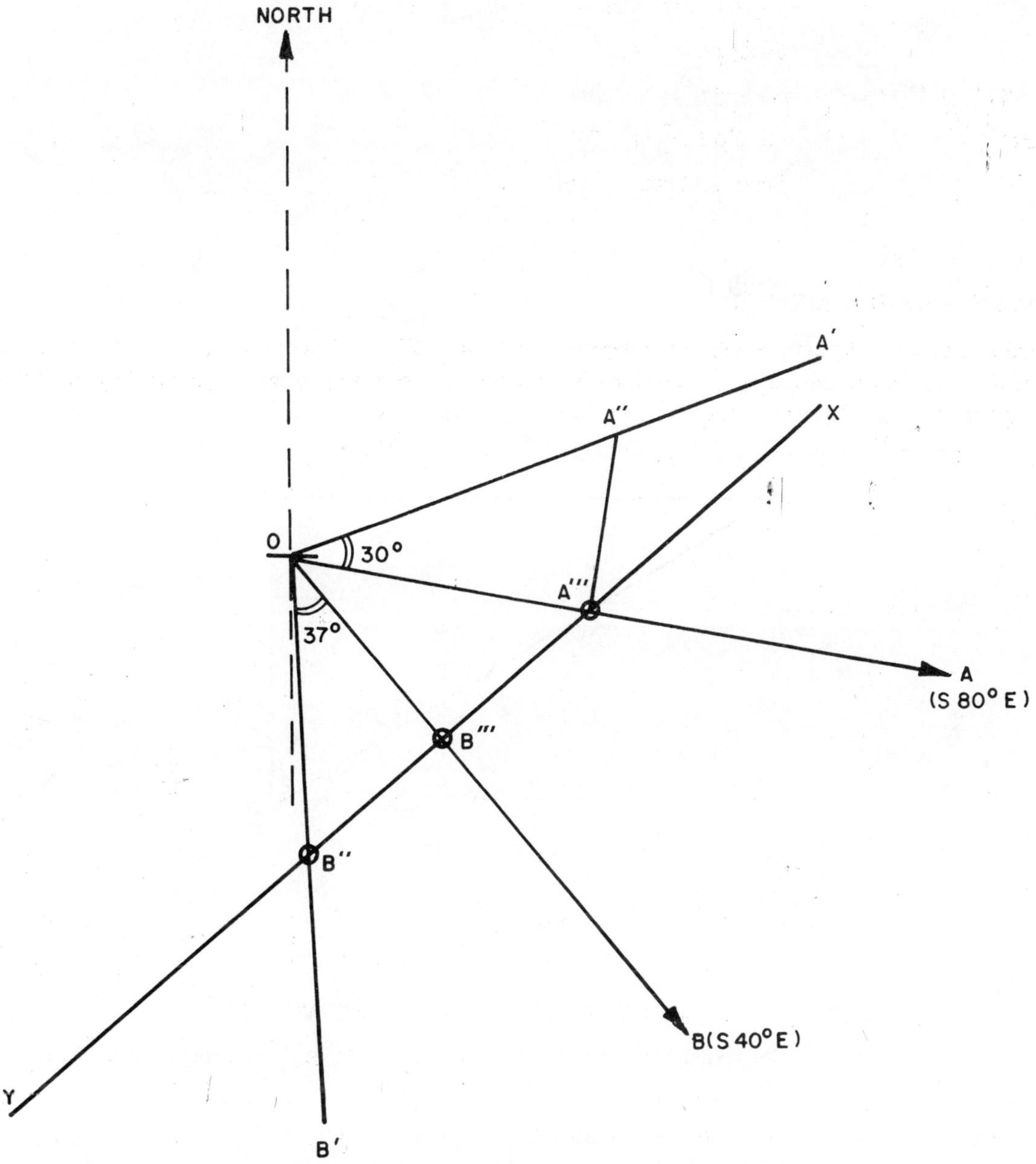

Fig. 1.1.7

Fig. 1.1.7 TRUE DIP AMOUNT AND DIRECTION, FROM ONE APPARENT DIP AND TRUE STRIKE (Problem 1.1.6)

OA	=	apparent dip direction, S 80 °E
AOA'	=	apparent dip amount (30°) along *OA*
A'''A''	=	perpendicular from *OA'* on *OA*
A'''A''	=	unit length= 2.54 cms
XA'''Y	=	true strike direction, N 50 °E – S 50 °W

OB	=	perpendicular from *O* on *XA''Y* = true dip direction = S 40 °E
B'''B''	=	unit length = 2.54 cms
BOB'	=	true dip amount = 37°

Scale 2.54 cms. = 1 unit

1.1.8 TANGENT VECTOR METHOD

This method is used when the dip is expressed in degrees like 20°,40°, 60° etc. The dip also can be stated as a gradient like 1 : 5, 1: 10, 1 : 25, etc. In the diagram given below, the relation between the gradient (meaning dip), the difference in elevation, and the horizontal distance between two points is clarified.

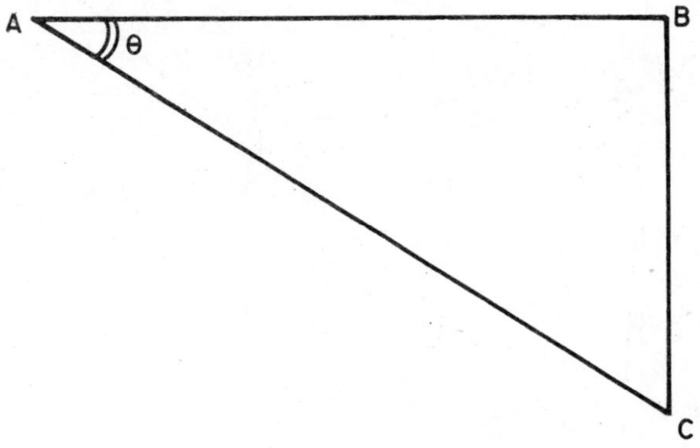

Fig. 1.1.8

Fig. 1.1.8 RELATION BETWEEN DIP AND TANGENT VECTOR OF DIP AMOUNT

AB	=	Horizontal distance between points *A* and *C*, these points being located on inclined surface *AC*.
theta	=	Dip amount of surface *AC*
BC	=	Difference between height (or depth) between points *A* and *C*
AC	=	slope or gradient line between points *A* and *C*

$$\tan\theta \text{ (dip amount)} = \frac{BC}{AB} = \frac{\text{depth (or height)}}{\text{horizontal distance}}$$

From the above figure (Fig. 1.1.8) it is clear that,

$$\tan \textit{theta} \text{ (amount of dip)} = \frac{BC}{AB}$$

Therefore the measure of the slope of line AC is automatically obtained by the value of the tangent of the concerned angle of dip. Thus the dip (θ) gets converted into a vector quantity ($\tan\theta$). Tangent of 45° is 1(one). Therefore choosing a convenient unit length, the tangent values of other angles (20°, 40°, 60°, 70°) can be reduced to appropriate lengths and shown along the concerned direction of (apparent) dip. This will be illustrated with examples given below.

1.1.8 (a) TRUE DIP FROM TWO APPARENT DIPS

Problem 1.1.7 A formation dips 20° due S 20 °W and 56° due S 40 °E. Find the amount and direction of true dip by the tangent vector method.

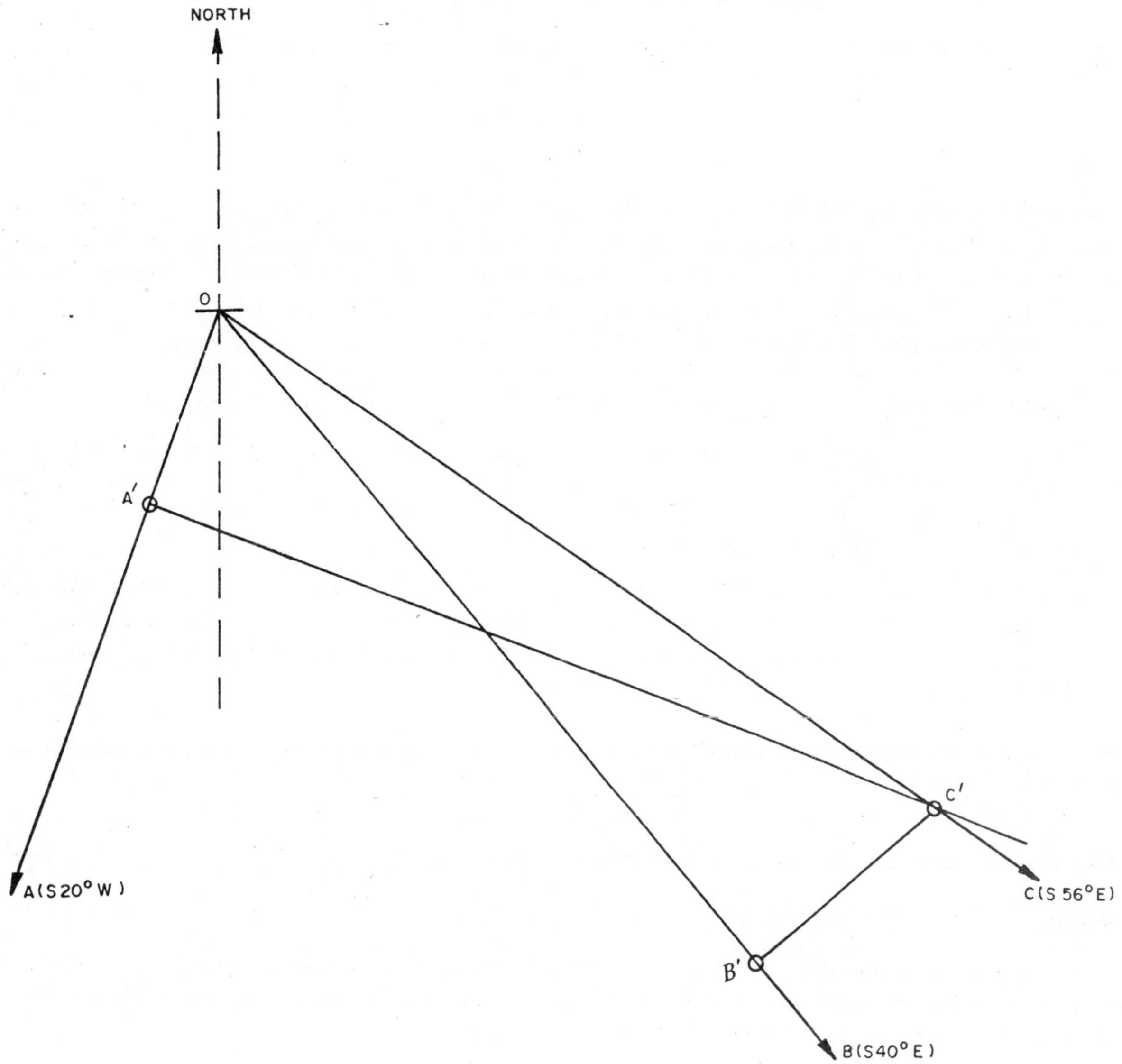

Fig. 1.1.9

Fig. 1.1.9 TRUE DIP FROM TWO APPARENT DIPS (Problem 1.1.7)

OA	=	apparent dip direction, S 20 °E
OA'	=	tangent vector of 20° = 0.364 (length is 3.64 cms)
A'C'	=	perpendicular at *A'* on *OA*
OB	=	apparent dip direction, S 40 °E
OB'	=	tangent vector of 50° = 1.482 (length is 14.82 cms)
B'C'	=	perpendicular at *B'* on *OB*
OC'C	=	true dip direction = S 56 °E
OC'	=	15.3 cms = tangent vector of value 1.53 (56° 48)

Note 1 cm = 0.1 value of tangent vector has been taken for convenience of plotting.

Procedure

From point *O*, draw lines *OA* and *OB* due S 20 °W and S 40 °E and mark off lengths 0.364 and 1.4826 corresponding to values of tangent for 20° and 56°, respectively, along these lines (Fig.1.1.9). Erect perpendiculars at *A'* and *B'* and extend them till these meet at point *C'*. Join points *O* and *C'* and extend this line to *C*. *OC'*C gives the direction of true dip and the distance *OC'* (1.53) gives the tangent vector amount of true dip. In the present case the amount of true dip is 57 degrees. The direction of true dip is S 56 °E.

1.1.8 (b) TRUE DIP AMOUNT, GIVEN, THE DIRECTION OF TRUE AND ONE APPARENT DIP

Problem 1.1.8 A formation has an apparent dip of 30° due south. If the true dip be due S 40 °W, find the amount of true dip by the tangent vector method.

Procedure

From point *O*, draw *OA* due south and mark off length *OA'* equal to tangent vector value of 30° (0.5774). Draw *OB* due S 40 °W (Fig. 1.2.0). Erect a perpendicular at *A'* and extend it to meet the line *OB* at *B'*. Measure length *OB'* which gives the tangent vector value of the true dip along OB. From this tangent vector value (0.75) read the value of angle of dip from the mathematical tables. It is found to be 36° 54'.

1.1.8 (c) TRUE DIP AMOUNT AND DIRECTION, GIVEN DIRECTION OF TRUE STRIKE AND ONE APPARENT DIP

Problem 1.1.9 A formation has an apparent dip of 35° due S 60 °E. If the true strike direction be N 60 °E - S 60 °W, find the direction and amount of true dip by the tangent vector method.

Procedure

In such cases though the direction of true strike has been given, the direction of true dip could be on any one side of the strike line. The direction gets fixed by the given direction of apparent dip which can be ascertained only by drawing the direction of the apparent dip. The procedure is described below.

From point *O* draw *OA* due S 60 °E and mark off tangent vector of 35° (0.7002) and obtain point A' (Fig. 1.2). Through *O* draw *COC'* trending N 60 °E – S60 °W, this being the given direction of true strike. It will be observed that the given direction of apparent dip is on the southern side of the strike line. Therefore draw *OB* perpendicular to the strike line, but on the southern side of it. Erect a perpendicular at *A'* on *OA* and extend it to meet *OB* at *B'*. *OB* gives the direction of true dip (S 30°E) and *OB'* gives the tangent vector value of the true dip due S 30 °E. The value is 0.8 which corresponds to 38° 42'.

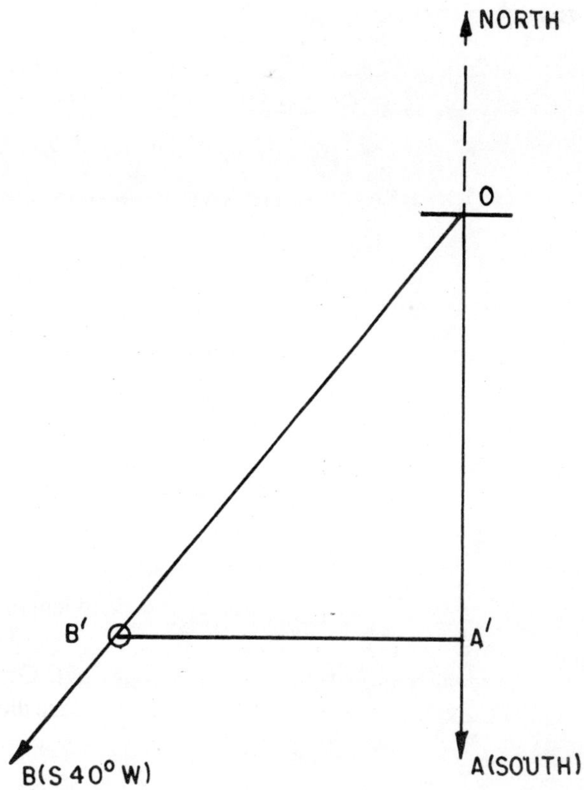

Fig. 1.2.0

Fig. 1.2.0 TRUE DIP AMOUNT, FROM ONE APPARENT DIP AND TRUE DIP DIRECTION (Problem 1.1.8)

OA	=	apparent dip direction, south
OA′	=	tangent vector of 30° = 0.5774 (length is 5.77 cms)
OB	=	true dip direction, S 40 °W
A′B′	=	perpendicular at *A′* on *OA*
OB′	=	7.5 cms tangent vector of value 0.75 = 36° 54 = true dip amount

Note 1 cm = 0.1 value of tangent vector for convenience of plotting.

In the above example (Problem 1.1.9), if the apparent dip be reversed to N 60 °W, the result obtained is different. It will be described further by giving another example.

Problem 1.2.0 A formation has an apparent dip of 35° due N 60 °W. If the direction of true strike be N 60 °E - S 60 °W, find the amount and direction of true dip.

Procedure

From point *O* draw *OA* due N 60 °W and mark off tangent vector of 35° (0.7002) on it to obtain point A′ (Fig. 1.2.2). Through *O* draw line *COC′* in N 60 °E - S 60 °W direction which is the given direction of true strike. Draw *OB* perpendicular on *COC′* at *O*, but on the northern side of it, because the given direction apparent dip is due N 60 °W and not due S 60 °E (as

1.1.9 DATA IN THE FORM OF DEPTH IN BOREHOLES

The apparent dip can be given in the form of depth of the formation ascertained in boreholes. Even this data can be set in the form of text, in the form of sketch, or set out in a tabular form. All such cases will be described in the following paragraphs.

1.1.9 (a) DATA GIVEN IN TEXT WITHOUT A SKETCH OF THE POSITION OF THE BOREHOLES

Problem 1.2.1 Boreholes are sunk to ascertain the attitude of a dipping body. Borehole A reaches the top of the body at a depth of 500 meters. Borehole B is 600 meters S 25 °W of borehole A, and it reaches the body at a depth of 200 meters. Borehole *C* is 1200 meters S 50 °E of borehole B, and it reaches the body at a depth of 600 meters. Draw a plan of the boreholes to a scale of 1 cm = 100 meters, and determine the attitude of the dipping body. The ground surface is flat.

Procedure

A plan of the boreholes is necessary because from the data given in the text, apparent dips along two directions can be directly calculated and the direction of true dip obtained. However it is advisable to consider three apparent dips on a dipping body. This is so because the accuracy of the calculations gets automatically checked when the three points of apparent dip fall or do not fall in a straight line. *Straightness of a line is defined by three points on it and not by two points.*

Step I Drawing a Plan of the Boreholes to the Given Scale

Plot the positions of the 3 boreholes to the scale of 1 cm = 100 meters and indicate the values of depth by the side of the concerned borehole, as shown in Fig. 1.2.3. Measure the distance between boreholes *A* and *C*, and also the bearing of such a line *(AC)*

Step II Calculation of Apparent Dips

The data in respect of the distance between the three boreholes, the bearing between the boreholes and the depth of the body in the three boreholes, are utilized to calculate the direction and the amount of apparent dip. If the ground surface be flat, then the direction of the apparent dip is from borehole of shallower level to one of the deeper level. In the present case (Problem 1.2.1) the three dips work out as shown below.

(a) apparent dip along *BA* i.e., due N 25 °E is equal to the distance between the two boreholes (600 meters) divided by the difference between the levels of the two boreholes (500 minus 200).

apparent dip along *BA* i.e.,

due N 25 °E $= \dfrac{600}{500-200} = 1:2$

(b) apparent dip *BC* i.e., due S 50 °E is equal to the distance between the two boreholes (1200 meters) divided by the difference between levels of the two boreholes (600 minus 200)

apparent dip along *BA* i.e.

due S 50 °E $= \dfrac{1200}{600-200} = 1:3$

(c) apparent dip along *AC* i.e., due S 27 °E is equal to the distance between the two boreholes (1470 meters) divided by the difference between the levels of the two boreholes (600 minus 500).

apparent dip along *AC* i.e.,

due S 27 °E $= \dfrac{1470}{600-500} = 1:14.7$

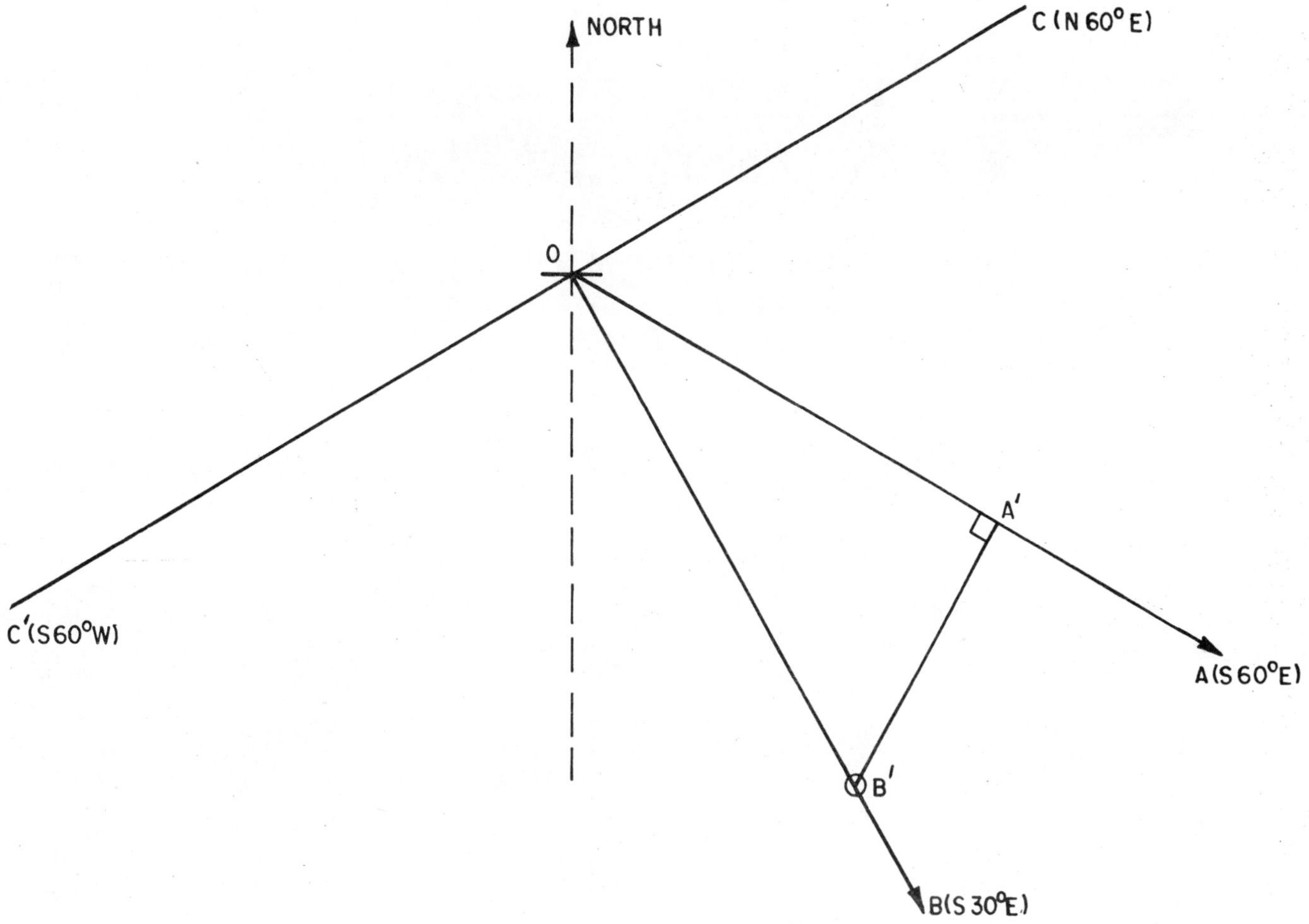

Fig. 1.2.1

Fig. 1.2.1 TRUE DIP DIRECTION AND AMOUNT, FROM ONE APPARENT DIP AND TRUE STRIKE DIRECTION (Problem 1.1.9)

OA	=	apparent dip direction S 60 °E
OA'	=	tangent vector of 35° = 0.7002 (length is 7.0 cms)
COC'	=	true strike line, N 60 °E – S 60 °W
OB	=	perpendicular at *O* on *COC'*= true dip direction = S 30 °E
A'B'	=	perpendicular at *A'* on *OA*
OB'	=	8.0 cms = tangent vector of true dip = 0.8 = 38° 42

Note 1 cm = 0.1 value of tangent vector has been taken for convenience of plotting.

was the case in Problem 1.1.9). Erect a perpendicular at A′ on *OA* and extend it to touch *OB* at *B′*. OB gives the direction of true dip as N 30 °W and *OB′* gives the tangent vector of the amount of the dip along *OB*. It is found to be 38° 42′. (0.8)

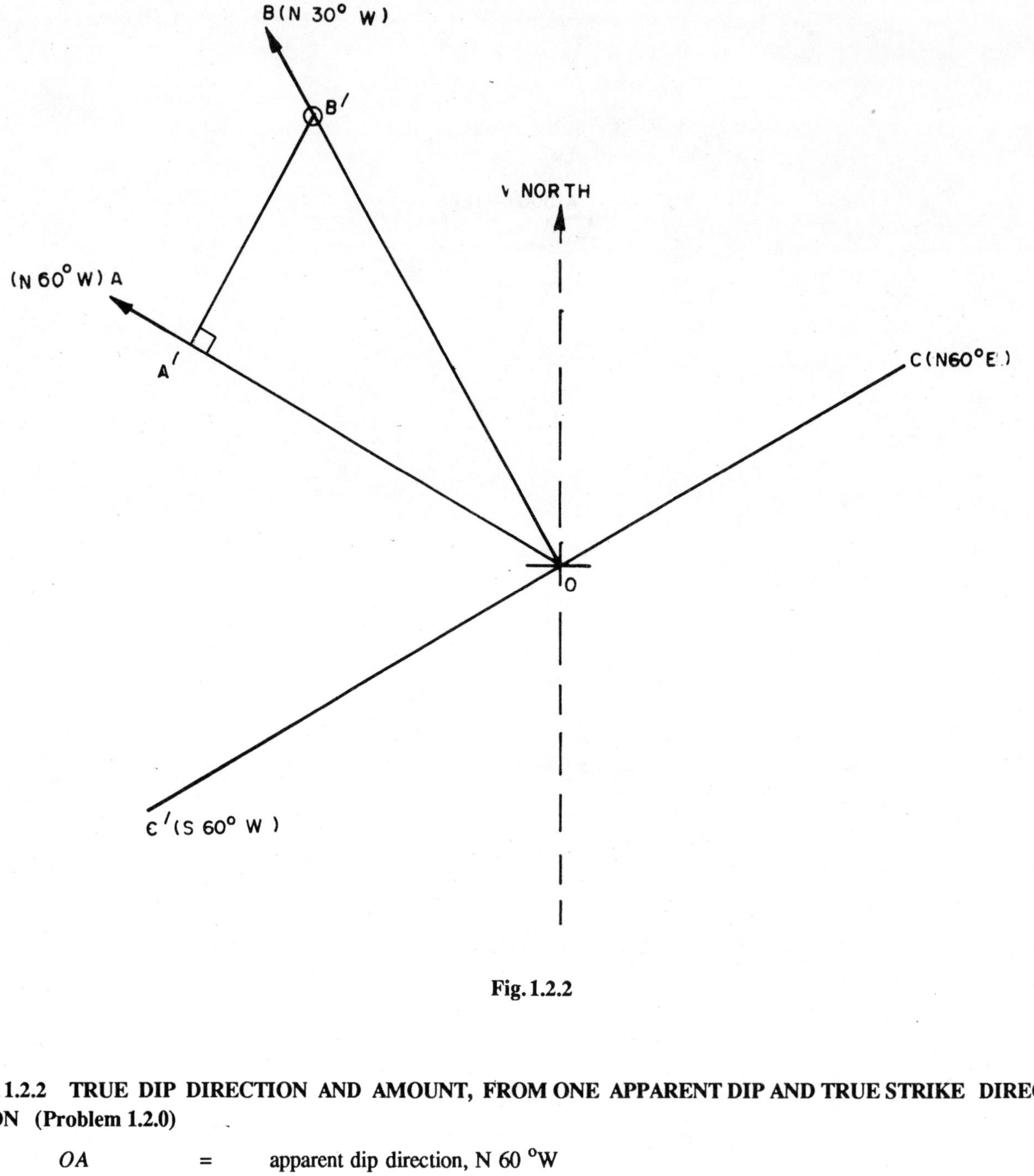

Fig. 1.2.2

Fig. 1.2.2 TRUE DIP DIRECTION AND AMOUNT, FROM ONE APPARENT DIP AND TRUE STRIKE DIRECTION (Problem 1.2.0)

OA	=	apparent dip direction, N 60 °W
OA'	=	tangent vector of 35° = 0.7002 (length is 7.0 cms).
COC'	=	true strike direction N 60 °E - S 60 °W
OB	=	perpendicular at *O* on *COC'* = true dip direction = N 30 °W
A'B'	=	perpendicular at *A'* on *OA*
OB'	=	8 cms = tangent vector of true dip = 0.8 = 38° 42

Note 1 cm = 0.1 value of tangent vector has been taken for convenience of plotting.

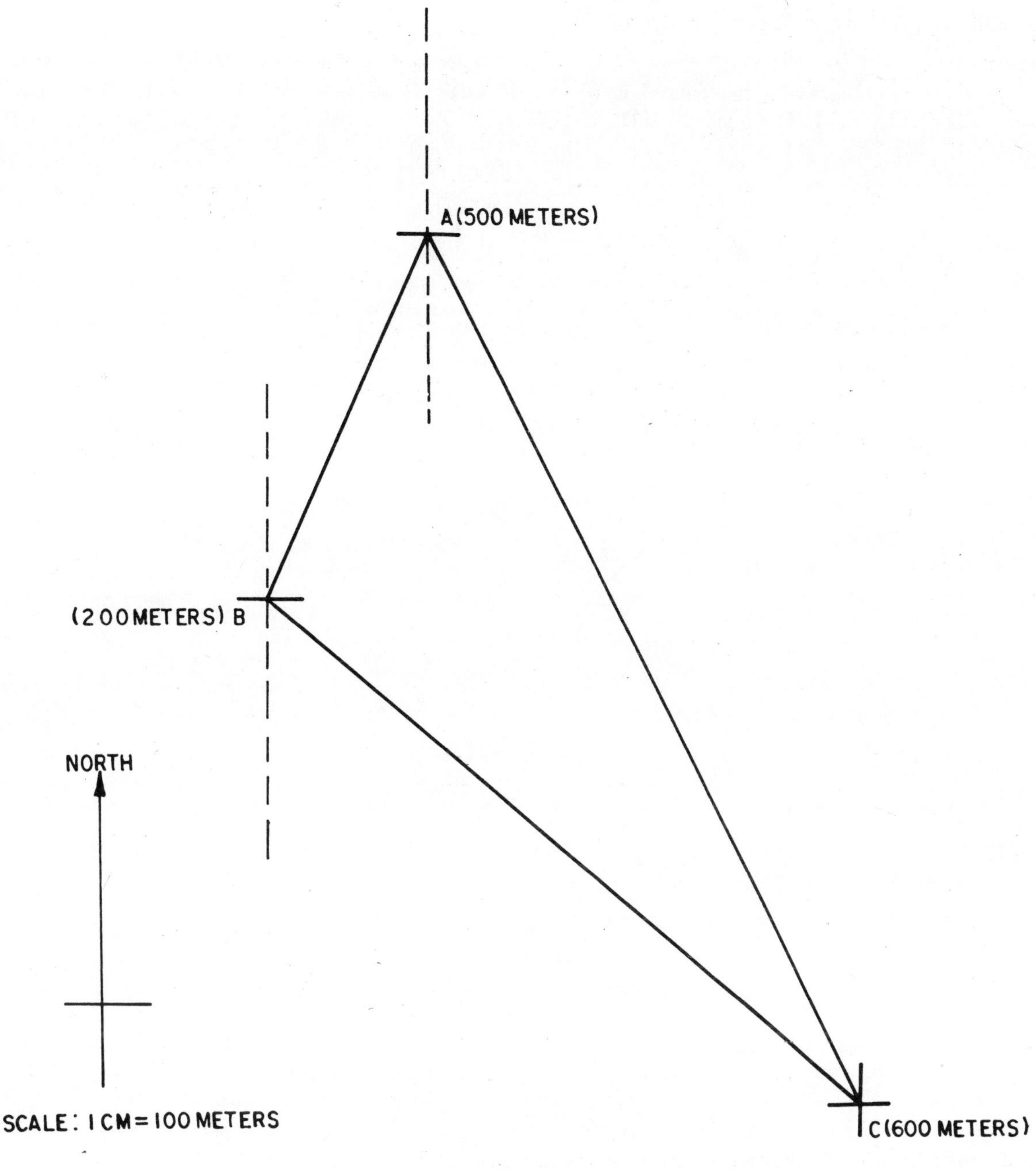

Fig. 1.2.3

Fig. 1.2.3 PLOTTING OF POSITIONS OF BORE HOLES A, B, C, (Problem 1.2.1)

Distance	**Direction/Bearing**
A - B, 600 meters	*AB*, S 25 °W
B - C 1200 meters	*BC*, S 50 °E
A - C 1470 meters (obained after plotting)	*AC*, S 27 °E (obtained after plotting)

Step III Plotting of Apparent Dips

A convenient unit should be taken to accomodate the values of apparent dips obtained in Step II above. Bigger the unit, greater is the accuracy of the result. From point *O* draws rays *OP, OQ* and *OR* due N 25°E, S 50°E and S 27°E, respectively. (Fig. 1.2.4). Mark off 2, 3 and 14.7 units along *OP, OQ* and *OR* respectively, these being the values of apparent dips, and obtain points.

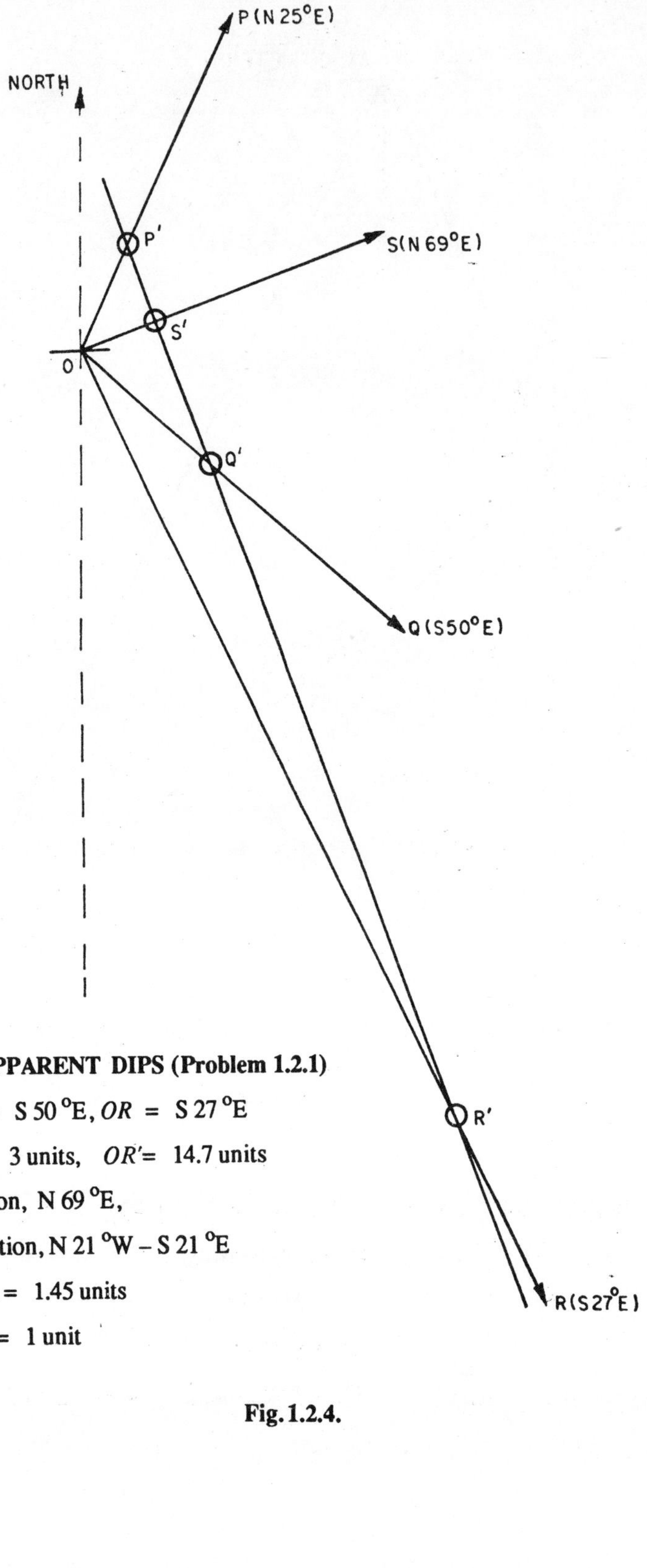

Fig. 1.2.4 PLOTTING OF APPARENT DIPS (Problem 1.2.1)

OP = N 25 °E, *OQ* = S 50 °E, *OR* = S 27 °E

OP'= 2 units, *OQ'* = 3 units, *OR'*= 14.7 units

OS = true dip direction, N 69 °E,

P'Q'R'= true strike direction, N 21 °W – S 21 °E

OS'= true dip amount = 1.45 units

Scale = 1 cm. = 1 unit

Fig. 1.2.4.

P', Q', and R'. Join P', Q' and R' by a straight line and drop a perpendicular OS from O, on the line drawn through points P', Q', and R'. The line $P'Q'R'$ gives the direction of true strike while the line OS gives the direction of true dip (N 69 °E). Measure the distance OS' to the same scale as that taken along OP', OQ' and OR'. The distance OS' gives the amount of true dip, which is found to be 1 : 1.45.

1.1.9 (b) DATA GIVEN IN THE FORM OF SKETCH OF BOREHOLES AND THE OTHER DATA IN TEXT

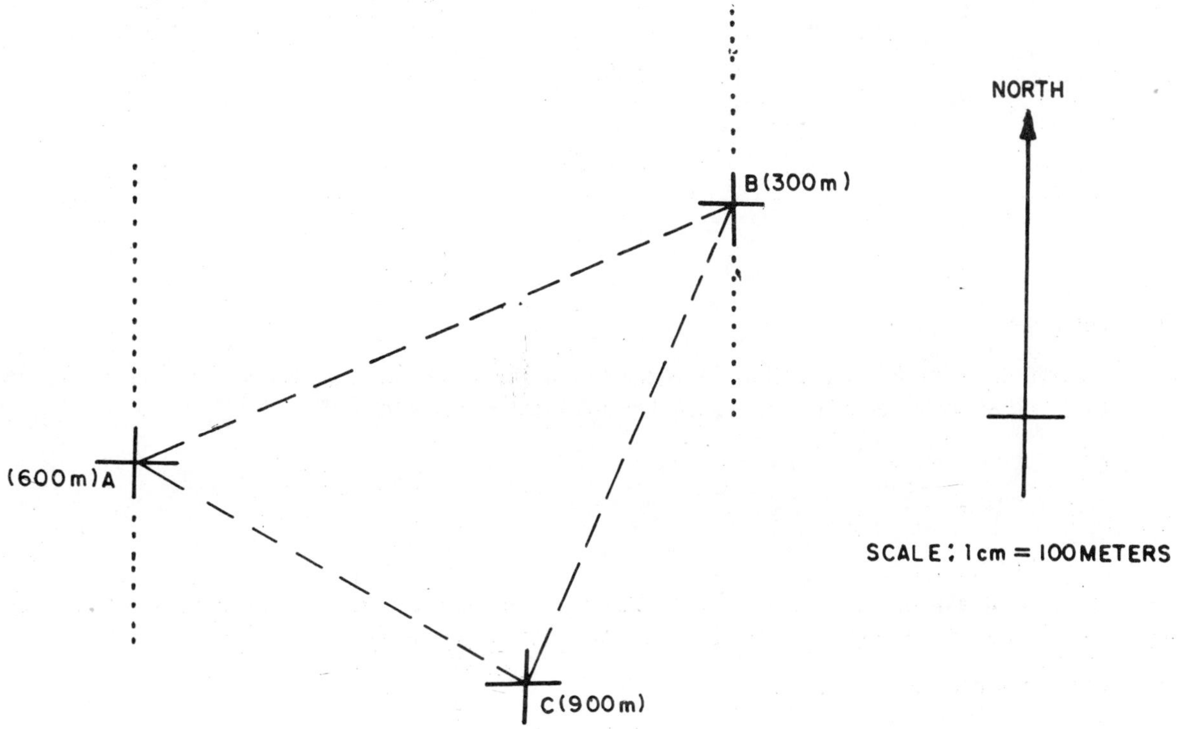

Fig. 1.2.5

Fig. 1.2.5. SKETCH OF BORE HOLES A, B, C, ON A SCALE OF 1 cm = 100 meters (Problem 1.2.2)

NOTE : In such cases the distances between the boreholes and the bearings between them are required to be measured from the sketch.

Problem 1.2.2. Boreholes are sunk at A, B and Fig. 1.2.5 and the top of a bedded formation is reached at depths of 600, 300 and 900 meters, respectively. Find the attitude of the bedded formation. The ground surface is flat.

Procedure

In such cases, the data in respect of the distance between the boreholes, the bearings between the boreholes are to be obtained by the actual measurement on the sketch. These data along with that of the depth at each borehole, are to be utilised to calclate the apparent dip directions and the amounts. However, in the present case it is not necessary to plot the data in the form of a sketch as it has been already presented in such a form (Fig. 1.2.5).

Step I Obtaining Data in Respect of Distance and the Bearing between the Boreholes

Between boreholes	*Distance in meters*	*Bearing*	*App. dip direction*	*Diff. in depths (in meters)*
A and *B*	900	*AB* N67 °E	*BA* S 67°W	300
B and *C*	740	*BC* S25 °W	*BC* S 25°W	600
A and *C*	600	*AC* S59°E	*AC* S59 °E	300

Step II Calculatioin of Apparent Dips

(a) apparent dip along *BA* i.e., due S 67 °W is equal to the distance between the boreholes A and B (900) divided by the difference in depth between the two boreholes (600 minus 300)

apparent dip along BA i.e.,

$$\text{due S } 67^{\circ}\text{ W} = \frac{900}{600-300} = 1:3$$

(b) apparent dip along *BC* i.e., due S 25 °W is equal to the distance between the boreholes *B* and *C* (740) divided by the difference in depth between the two boreholes (900 minus 300)

apparent dip along *BC* i.e,

$$\text{due S } 25^{\circ}\text{W} = \frac{740}{900-300} = 1:1.23$$

(c) apparent dip along *AC* i.e., due S 59 °E is equal to the distance between the boreholes A and C (600) divided by the difference in depth between the two boreholes (900 minus 600)

apparent dip along *AC* i.e.

$$\text{due S } 59^{\circ}\text{E} = \frac{600}{900-600} = 1:2$$

Step III Plotting of the Apparent Dips

From point *O* draw lines *OP*, *OQ* and *OR* due S 67 °W, S 25 °W and S 59 °E respectively. Mark off amounts of apparent dips 3, 1.23 and 2 along *OP*, *OQ* and *OR*, respectively, and obtain points *P′*, *Q′* and *R′*, respectively. Join points *P′ Q′* and *R′* by a straight line which gives the direction of true strike line. Drop a perpendicular *OS* on the strike line *P′Q′R′* from point *O*, which gives the direction of true dip. The dip direction is found to be due south and the amount of dip to be 1 : 1.5 (Fig. 1.2.6).

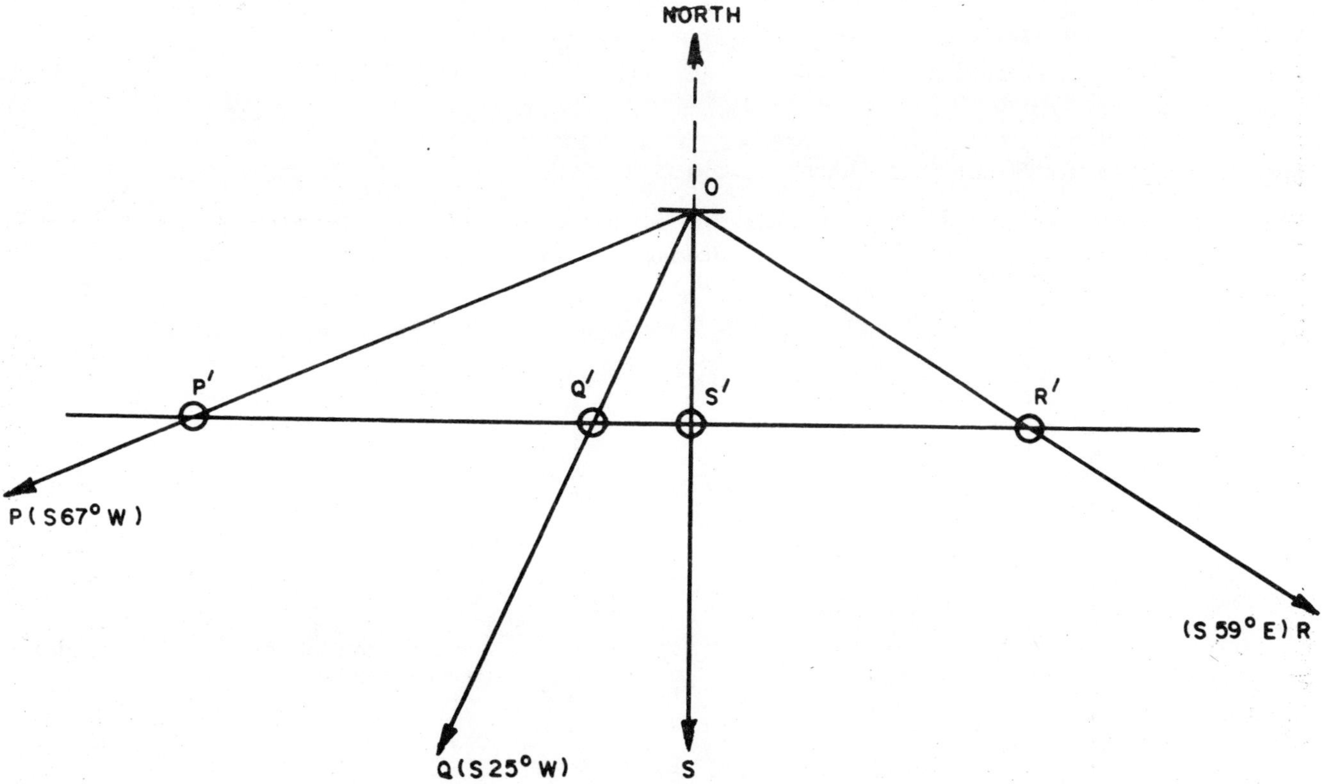

Fig. 1.2.6

Fig.1.2.6 PLOTTING OF APPARENT DIPS (Problem 1.2.2)

OP = S 67 °W, OQ = S 25 °W, OR = S 59 °E
OP' = 3 units, OQ' = 1.23 units, OR' = 2 units

OS = true dip direction, South, $P'Q'R'$ = true strike direction = E-W

OS' = true dip amount = 1.15 units

Scale 1 inch = 1 unit

1.1.9 (c) DATA GIVEN IN THE FORM OF TABULATION

Problem 1.2.3 Boreholes are sunk to ascertain the attitude of a tabular body. The data are given below.

Between boreholes	*Distance in meters*	*Bearing*	*App. dip direction*	*Diff. in depths (in meters)*
A and *B*	900	*AB* N67 °E	*BA* S 67°W	300
B and *C*	740	*BC* S25 °W	*BC* S 25°W	600
A and *C*	600	*AC* S59°E	*AC* S59 °E	300

Draw a plan of the boreholes to a scale of 1 cm = 100 meters. Determine the attitude of the tabular body. The ground surface is flat.

Procedure

A plan of the boreholes is needed to be drawn because only then the distance and the bearing between boreholes A and C can be determined.

Step I Plan of the Boreholes on a Scale of 1 cm = 100 meters

Draw a plan of the positions of the boreholes and measure the distance between the boreholes *A* and *C*, and also the bearing between these two boreholes.

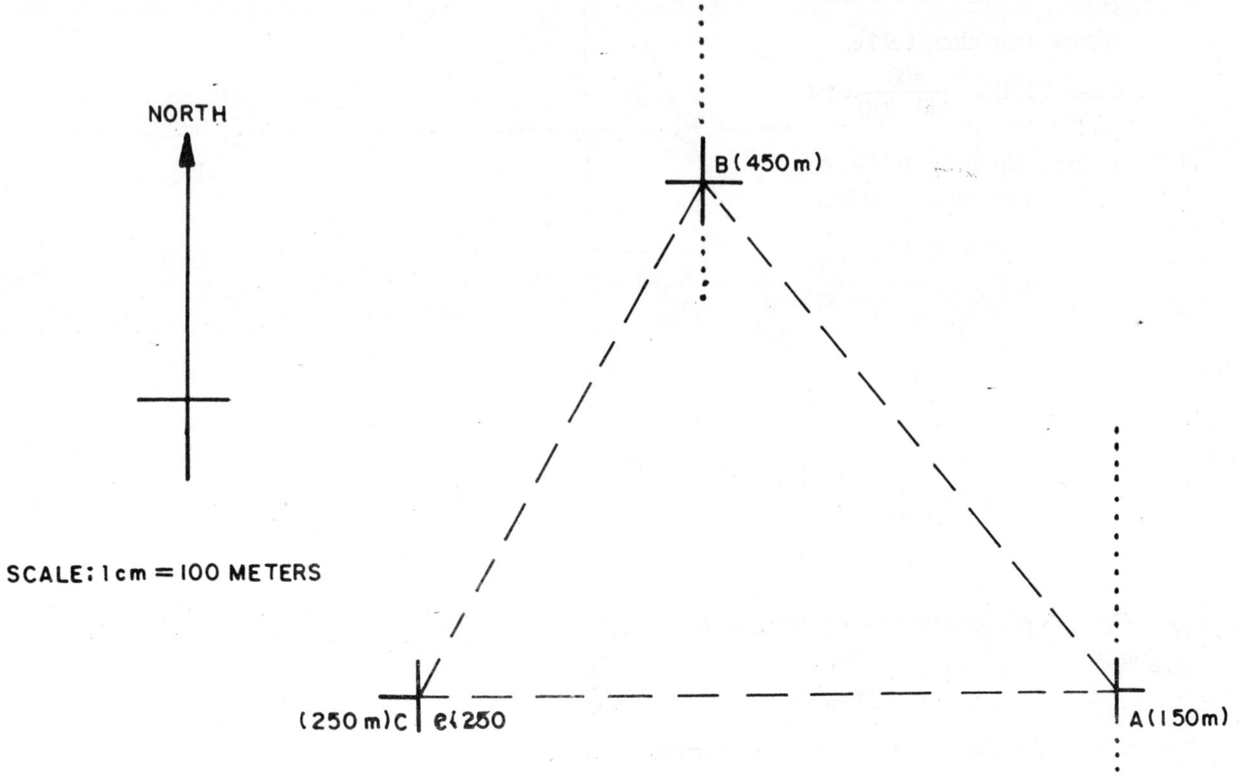

Fig. 1.2.7.

Fig. 1.2.7. PLOTTING POSITIONS OF BOREHOLES ***A, B, C*** **ON A SCALE OF 1 cm = 100 Meters. (Problem 1.2.3)**

NOTE : In such cases the bearing and the distance between boreholes *A* and *C* is required to be measured from the plotting of the data.

Step II Tabulation of Data as Derived from Step I

Between Boreholes	Distance and bearing	Difference in depth and direction of apparent dip
A and *B*,	900 meters *AB* N 40° W,	300 meters *AB* N 40° W
B and *C*,	800 meters *BC* S 30° W,	200 meters *CB* N 30° E
A and *C*,	980 meters *AC* S 89° W,	100 meters *AC* S 89° W

Step III Calculation of Apparent Dips

(a) apparent dip along *AB* i.e., due N 40 °W is equal to the distance between the boreholes *A* and *B* (900) divided by the difference in level between the two boreholes (450 minus 150).

apparent dip along *AB* i.e.,

$$\text{due N } 40\,^{\circ}\text{W} = \frac{900}{450-150} = 1{:}3$$

(b) apparent dip along *CB* i.e., due N 30 °E is equal to the distance between the boreholes *B* and *C* (800) divided by the difference in level between the two boreholes (450 minus 250).

apparent dip along *CB* i.e.,

$$\text{due N } 30\,^{\circ}\text{E} = \frac{800}{450-250} = 1{:}4$$

(c) apparent dip along *AC* i.e., due S 89 °W is equal to the distance between the boreholes *A* and *C* (980) divided by the difference in level between the boreholes. (250 minus 150).

apparent dip along *AC* i.e.,

$$\text{due S } 89\,^{\circ}\text{W} = \frac{980}{250-150} = 1{:}9.8$$

Step IV Plotting of the Apparent Dips

From point *O* draw *OP* *OQ* and *OR* due S 89 °W, N 40 °W and N 30 °E corresponding to *AC*, *AB* and *CB*, respectively. (Fig. 1.2.8.)

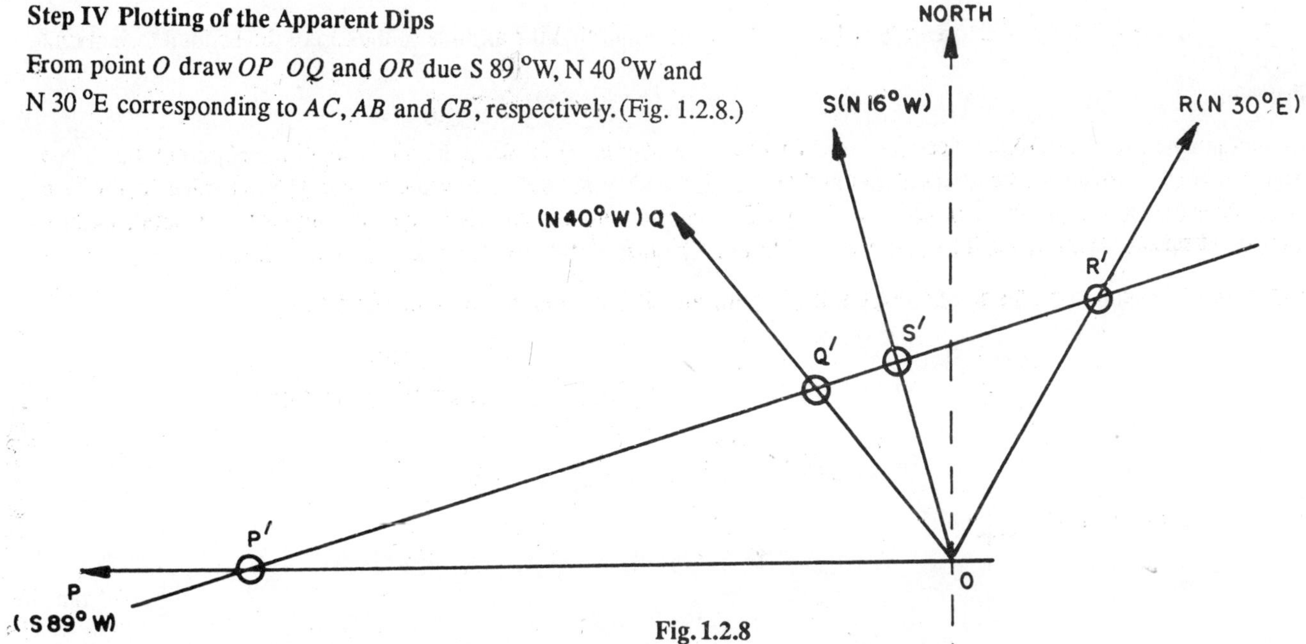

Fig. 1.2.8

Fig. 1.2.8. PLOTTING OF APPARENT DIPS (Problem 1.2.3)

OP = S 89 °W, *OQ* = N 40 °W, *OR* = N 30 °E

OP′ = 9.6 units, *OQ′* = 3 units, *OR′* = units

OS = true dip direction, N 16 °W,

P′Q′R′ = true strike direction, N 74 °E – S 74 °W

OS′ = true dip amount = 2.7 units

Scale 1 cm = 1 unit

Mark off apparent dip amounts 9.8, 3 and 4 along OP, OQ and , OR and obtain points P', Q' and R' respectively. Join points P', Q' and R' by a straight line which gives the true strike line. Drop a perpendicular OS on the strike line P' Q' R' cutting it at S'. OS gives the direction of true dip and the distance OS' gives the amount of true dip. The true dip direction and amount are found to be N 16°W and 1 : 2.7 respectively (Fig. 1.2.8).

1.1.9 (d) DATA GIVEN IN THE FORM OF TABULATION, BUT HEIGHTS ARE GIVEN INSTEAD OF DEPTH

So far data pertaining to boreholes have been described. It is possible however that a tabular body may crop out at the surface, but at different heights. If it corresponds to a bedding plane, then the attitude can be derived. The procedure adopted in such cases is different. It will be described below.

Problem 1.2.4 The top of a bedded formation is cropping out at three places, the details of which are as given below.

Place,	Location and Bearing,	Heights (meters)
A,	Nil,	500
B,	300 meters South of *A*,	600
C,	800 meters N 50°E of *B*,	400

Draw a plan of the places to a scale of 1 cm = 100 meters and find the attitude of the top of the bedded formation.

Procedure

It is not much different from that described in earlier cases, except in establishing the direction of the apparent dip. Since values of heights are given, the direction of dip therefore is from a point of higher value to that of lower value (in the case of borehole data, this is just the opposite). Further the distance between the spots is inclined one and not horizontal, because the ground surface is not flat and heights are considered. The rest of the procedure is described below.

Step I Drawing plan of places and finding out the bearing and distance between points *A* and *C*.

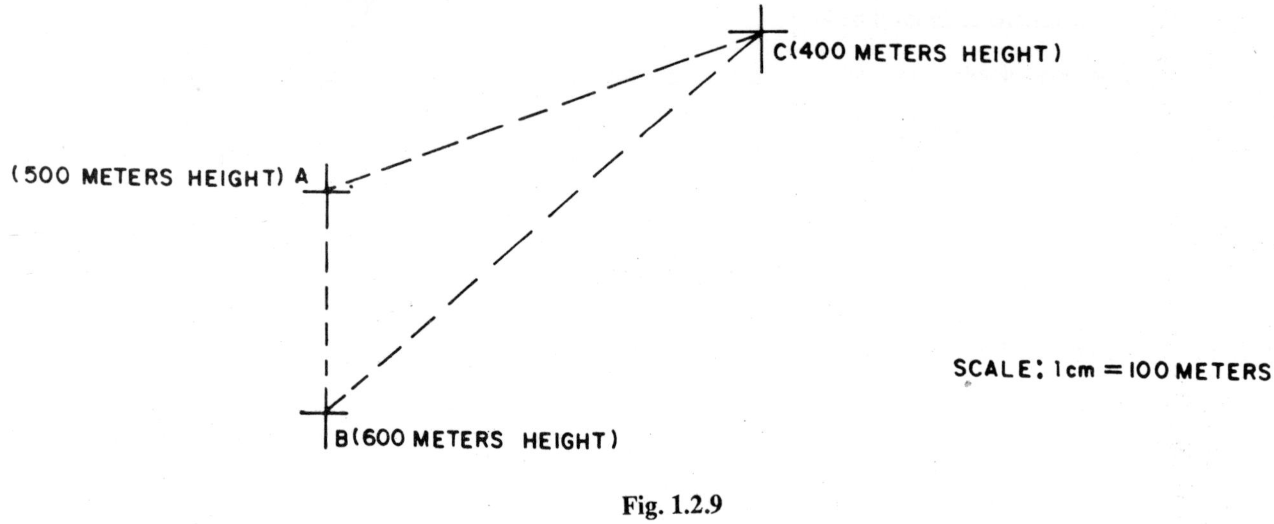

Fig. 1.2.9

Fig. 1.2.9. PLOTTING POSITIONS OF POINTS *A*, *B*, *C* ON A SCALE OF 1 cm = 100 Meters. (Problem 1.2.4)

NOTE : In such cases bearing and distance between points *A* and *C* is required to be measured from the plotting of the data.

Step II **Tabulation of Data Derived From Fig. 1.2.9**

Between places	*Distance and bearing*	*Difference in heights (meters)*	*Apparent Dip direction*
A and B	300 meters, AB, South	100	BA, North
B and C	800 meters, BC, N 50° E	200	BC, N50° E
A and C	650 meters AC, N 71° E	100	AC N 71° E

Step III **Calculation of Apparent Dips**

(a) apparent dip along *BA* i.e., due north is equal to the distance between the places *A* and *B* (300 meters) divided by the difference in level between the two places (600 minus 500).

apparent dip along *BA* i.e.,

due north $= \dfrac{300}{600-500} = 1:3$

(b) apparent dip along *BC* i.e., due N 50 °E is equal to the distance between the places *B* and *C* (800 meters) divided by the difference in level between the two places (600 minus 400)

apparent dip along *BC* i.e.,

due N 50° E $= \dfrac{800}{600-400} = 1:4$

(c) apparent dip along *AC* i.e., due N 71 °E is equal to the distance between the places *A* and *C* (650 meters) divided by the difference in level between the two places (500 minus 400).

apparent dip along *AC* i.e.,

due N 71 °E $= \dfrac{650}{500-400} = 1:6.5$

Step V Plotting of the Apparent Dips

From point *O* draw *OP*, *OQ* and due north, N 50 °E and N 71 °E respectively which correspond to apparent dip directions *BA*, *BC* and *AC*, respectively. Mark off apparent dip amounts 3, 4 and 6.5 along *OP*, *OQ* and *OR* and obtain points *P′ Q′* and *R′*, respectively. Join points *P′ Q′* and R′ by a straight line which gives the direction of true strike line. Drop a perpendicular *OS* on *P′*Q, Q^T, *R′* from *O*, which gives the direction of true dip. This line intersects the strike line at *S′*. Distance *OS′* measured to the same unit length as done for *OP′*, *OQ′* and *OR′*, gives the amount of true dip. It is fround to be 1 : 2.9, and the direction to be N 8 °E (Fig.1.3.0).

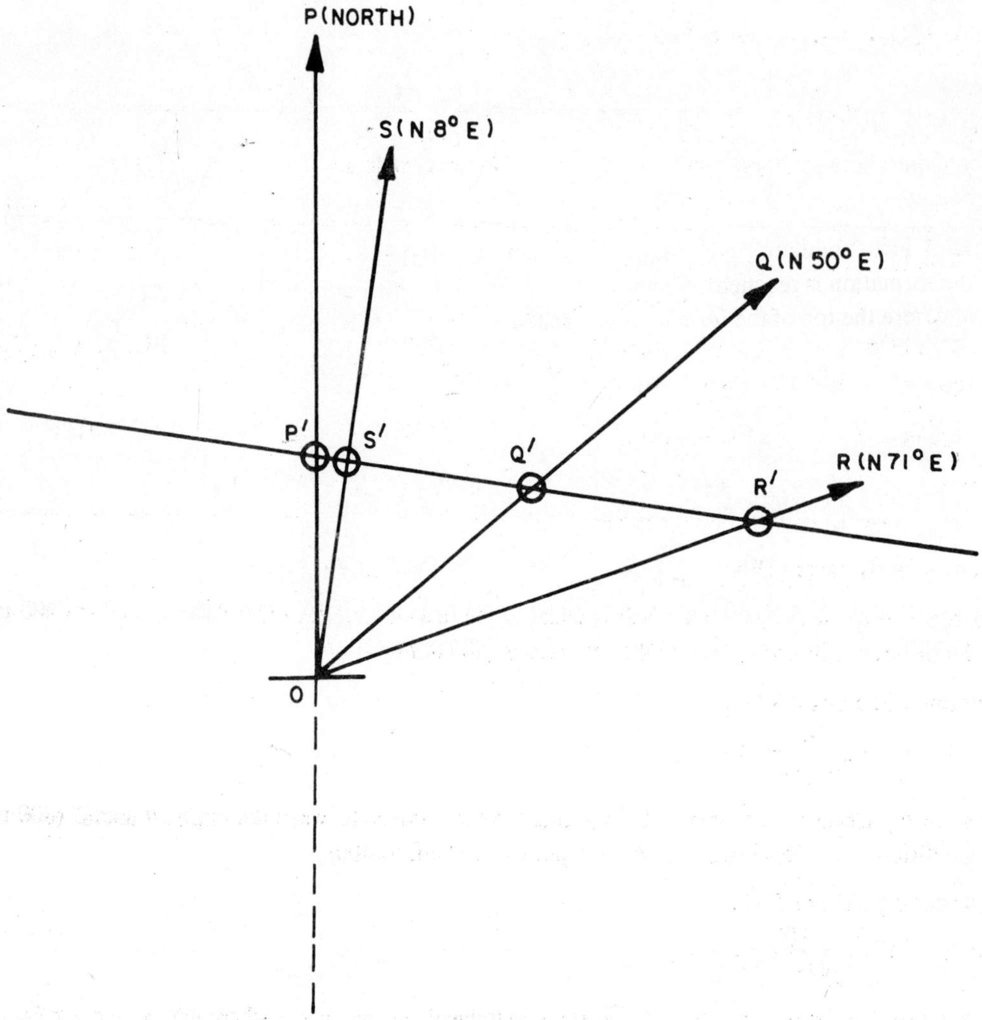

Fig. 1.3.0

Fig. 1.3.0. PLOTTING OF APPARENT DIPS (Problem 1.2.4)

OP = north, *OQ* = N 50 °E, *OR* = N 71 °E

OP' = 3 units, *OQ'* = 4 units, *OR'* = 6.5

OS = true dip direction, N 8 °E

P'Q'R' = true strike direction, N 82 °W – S 82 °E

OS' = true dip amount = 2.9 units

Scale 1 cm. = 1 unit

1.1.9 (e) BOREHOLE DATA FROM UNDULATING GROUND SURFACE

The last variety of setting the data comprises of boreholes sunk in an undulating land surface. This situation is the most common one. The actual depths recorded in the boreholes cannot be directly used to derive the direction of apparent dip, because the depths reached are not from any common horizon or level. Each borehole is at different topographical height and therefore the direct value of depth reached in it, cannot be used for the purposes of calculating the direction of dip, or the difference in depths between the boreholes. The procedure adopted will be described by considering an actual example which is given below.

Problem 1.2.5 In an undulating region, boreholes are sunk in three places. Borehole *A* has an elevation of 500 meters and the top of the formation is reached at a depth of 200 meters. Borehole *B* is 600 meters S 50 °E of *A*, has an elevation of 580 meters, and here the top of the formation is reached at a depth of 780 meters. Borehole *C* is 600 meters S 40 °W of *B*, has an elevation of 400 meters, and the top of the formation is reached at a depth of 300 meters. Draw a plan of the boreholes to a scale of 1 cm = 100 meters, and determine the attitude of the formation.

Procedure

As the boreholes are sunk on an undulating ground surface, the actual height of the borehole affects the value of the depth reached in the boreholes. Therefore the true values of depths in the three boreholes are calculated first and these latter values are then used to determine the direction and the amount of true dip.

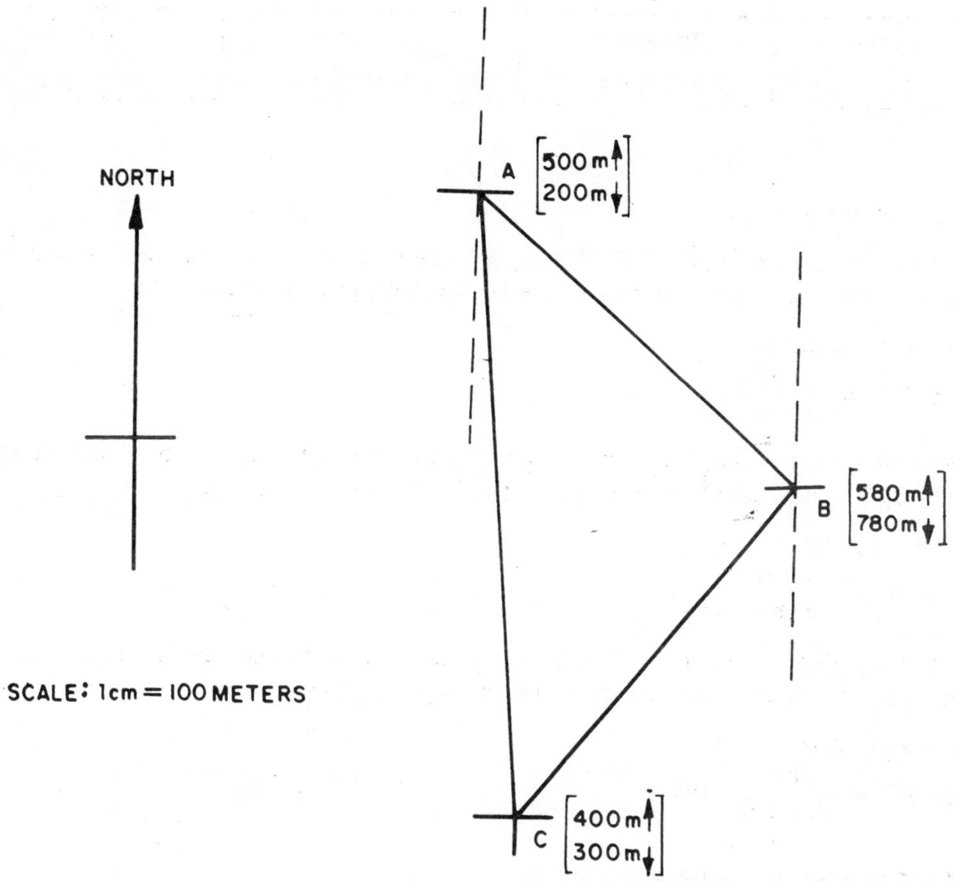

Fig. 1.3.1.

Fig. 1.3.1. PLOTTING OF POSITIONS OF BOREHOLES *A, B, C,* **ON A SCALE OF 1 cm. = 100 meters (Problem 1.2.5)**

NOTE : In such cases the bearing and the distance between boreholes *A* and *C* is required to be measured from the plotting of the data.

Step I Drawing plan of boreholes to the scale of 1 cm = 100 meters. Indicate the height by arrow pointing upwards and depth by arrow pointing downwards. Measure the distance between the boreholes *A* and *C*, and calculate the bearing between them from the plan.

Step II **Tabulation of data derived from fig. 1.3.1**

Between boreholes	*Distance and bearing*	*Value of actual depth in boreholes (meters)*	*Apparent Dip direction*
A and B	600 meters, AB, S 55 ° E	A 300 (+) B 200 (−)	AB, S 50 ° E
B and C	600 meters, CB, S 40° W	B 200 (−) C 100 (+)	CB, N 40° E
A and C	850 meters AC, S 5° E	A 300 (+) C 100 (+)	AC S 5° E

Step III Calculation of Apparent Dips

(a) apparent dip along *AB* i.e., due S 50 °E is equal to the distance between the boreholes *A* and *B* (600 meters) divided by the difference in level between the two boreholes (300 minus minus 200)

apparent dip along *AB* i.e.,

$$\text{due S } 50\,^{\circ}\text{E} = \frac{600}{300-(-200)} = 1{:}1.2$$

(b) apparent dip along *CB* i.e., due N 40 °E is equal to the distance between the boreholes *B* and *C* (600 meters) divided by the difference in level between the two boreholes (100 minus minus 200).

apparent dip along *CB* i.e.,

$$\text{due N } 40\,^{\circ}\text{E} = \frac{600}{100-(-200)} = 1{:}2$$

(c) apparent dip along *AC* i.e., due S 5 °E is equal to the distance between the boreholes *A* and *C* (850 meters) divided by the difference in level between the two boreholes (300 minus 100).

apparent dip along *AC* i.e.,

$$\text{due S } 5\,^{\circ}\text{E} = \frac{850}{300-100} = 1{:}4.25$$

PLOTTING OF THE APPARENT DIPS

Step IV

From point *O* draw *OP*, *OQ* and *OR* due NJ 40°E, S50°E and S5°E corresponding to *CB*, *AB* and *AC*, respectively. Mark off apparent dip amounts of 2.1.2 and 4.25 along *OP*, *OQ*, and *OR*, respectively, and obtain points *P'*, *Q'* and R'.....by a straight line. Draw a perpendicular *OS* on it from *O*. *OS* give the direction of true dip. It cuts the strike line at point S'. Distance *OS'* gives the amount of true dip. The amount of true dip is found to be 1:1.05 and the direction to be S 81°E.

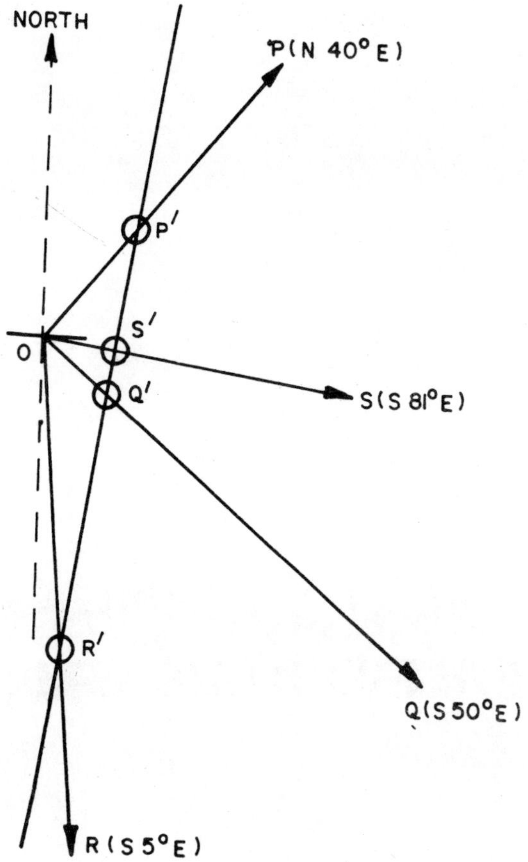

Fig. 1.3.2

Fig. 1.3.2 PLOTTING OF APPARENT DIPS (Problem 1.2.5)

OP = N 40 °E, OQ = S 50 °E, OR = S 5 °E

OP' = 2 units, OQ' = 1.2 units, OR' = 4.25 units

OS = true dip direction, S 81 °E

$P'Q'R'$ = true strike direction, N 9 °E – S 9° W

OS' = true dip amount = 1.05 units

Scale 1 cm. = 1 unit

In the preceeding pages of this chapter, various modes of setting the data of apparent dip, and the procedure to be adopted in determining the direction of true dip and the strike, has been elaborated. Once the direction of true dip is obtained, then the amount of apparent dip in any other required direction, or the direction in which a particular amount of apparent dip can be encountered, is easily derivable, the procedure for which also has been described.

2
ESTABLISHMENT OF KIND OF STRUCTURE OF GEOLOGIC BODY

2.1.0 INTRODUCTION

The knowledge about the direction and the amount of dip is utilised to establish the kind of the structure possessed by the geologic body. Structures like fold, fault etc., can be inferred only from the behaviour of the direction of the dip, the difference between the expected level and the actual level of a particular horizon in the case of a fault, and so on. This is possible only if the given data are carefully analysed. Some cases will be described below.

2.1.1. Case I

EXISTENCE OF A GEOLOGIC BODY POSSESSING ONLY ONE DIRECTION OF INCLINATION

In such cases, the data may be set out in the form of text, tabulation or a sketch of the boreholes. Three boreholes are enough. Most of the problems given in Chapter I yield the result that the bedded formation is having only one direction of true dip and the amount of true dip is constant. This is required to be confirmed by obtaining borehole data from new set of boreholes from the same area. Geologic body possessing uniform dip is a very simple structure and its existence is only theoretically possible. In actuality, the true dip is not constant when studied over a considerable area. In this book the constancy of the dip direction and the amount is presumed to avoid complications. Wherever only three boreholes are considered, it is reasonable to infer that the geologic body possesses simple structure of uniformly dipping rocks. Thus the geologic structure derived for the problems 1.1.0 to 1.1.2, 1.1.4 to 1.1.2 is as given below (the problems are given in Chapter I).

Problem No.	Kind of Structure
1.1.0	A dipping sandstone with a dip of 1:4.2 due S 53 °E.
1.1.1.	Bedded formation dips 1:4.2 due N 50 °E
1.1.2	It has true dip due N 55° W and also due S 7 °E, but it is not a fold. It is so because only one apparent dip has been given, which therefore has yielded two directions of true dip.

1.1.4 Formation dips 58° due S 53°W

1.1.5 It is dipping formation with true dip of 42° due N 50 °E

1.1.6 It is dipping formation with a true dip of 37° due S 40 °E

1.1.7 The body is dipping with a true dip of 57° due S 56 °E

1.1.8 It is dipping body with a true dip of 36° 54′ due S 40°W

1.1.9 It is dipping body with a true dip of 38° 42′ due S 30 °E.

1.2.0 It is dipping body with a true dip of 38° 42′ due N 30 °W

1.2.1 It is dipping body with a true dip of 1 : 1.45 due N 69 °E

1.2.2 It is dipping body with a true dip of 1 : 1.15 due South.

1.2.3. It is dipping body with a true dip of 1 : 2.7 due N 16 °W

1.2.4 It is dipping body with a true dip of 1 : 2.9 due N 8 °E.

1.2.5 It is dipping body with a true dip of 1 : 1.05 due S 81 °E.

2.1.1 Case II A Folded Geologic Body

A. Anticlinal Fold

Problem 2.1.0 Boreholes are sunk to ascertain the structure of a folded body. Two sets of boreholes are sunk, each corresponding to the two limbs of the fold.

SETI I Borehole *A* reaches the body at a depth of 500 meters. Borehole *B* is 900 meters west of *A* and it reaches the body at a depth of 800 meters. Borehole *C* is 400 meters S 50°E of B and it reaches the body at a depth of 600 meters.

SET II These boreholes are sunk to the immediate east of borehole *A*. Borehole *P* reaches the same body at a depth of 600 meters. Borehole *Q* is 900 meters east of *P* and it reaches the body at a depth of 900 meters. Borehole *R* is 400 meters N 50 °W of *Q*, and it reaches the body at a depth of 700 meters.

The ground surface is flat. Draw a plan of the boreholes to a scale of 1 cm = 100 meters and determine the structure of the geologic body.

Procedure

The data set out for the two sets should be dealt with separately. The true dip direction and the amount for the two sets should be determined first. The result so obtained is to be interpreted to derive the structure.

Step I Drawing the Plan of the Two Sets of Boreholes
(Not to scale)

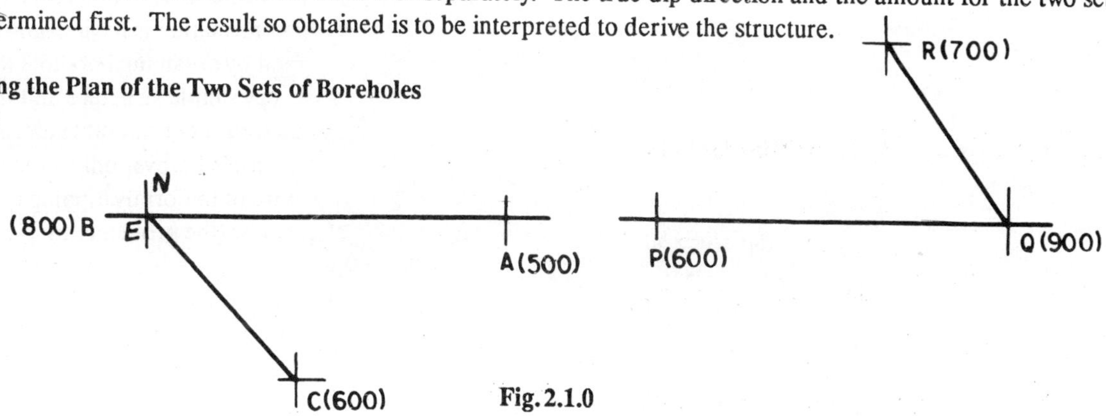

Fig. 2.1.0

Fig. 2.1.0. PLAN OF BORE HOLES *A, B, C* AND *P, Q, R*, (NOT TO SCALE) (Problem 2.1.0)

NOTE : This helps to plot the data accurately.

Step II Determining the Direction and Amount of Apparent Dips, and Deriving the Amount and Direction of True Dip for the two Sets.

SET I (a) Plotting of Boreholes A, B and C to a scale of I cm = 100 meters and Collection of Data regarding Distances, Bearing Between Boreholes (Fig. 2.1.1)

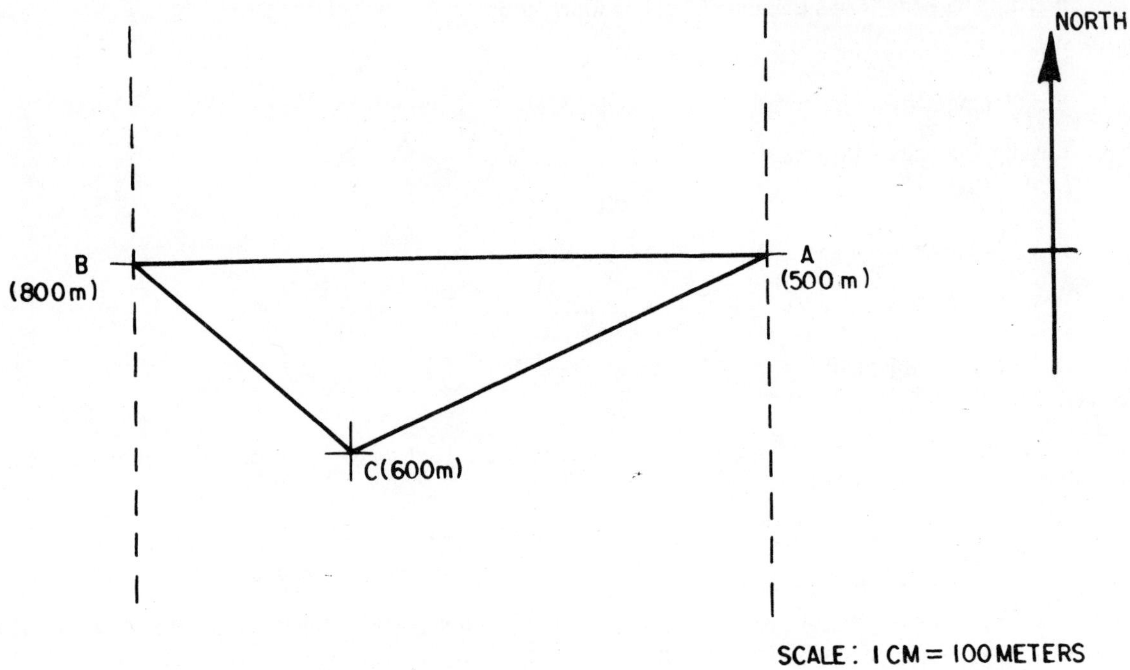

Fig. 2.1.1.

Fig. 2.1.1. Plotting of Positions of Bore Holes *A, B, C,* to a Scale of 1 cm. = 100 meters (Problem 2.1)

NOTE In such cases, the bearing and distance between boreholes *A* and *C* is to be measured from the plotting of the data.

(b) Calculation of Apparent Dips for Set I

(i) apparent dip along *AB* i.e.,

$$\text{due west} = \frac{900}{(800-500)} = 1{:}3$$

(ii) apparent dip along *CB* i.e.,

$$\text{due N } 50^\circ \text{ W} = \frac{400}{(800-600)} = 1{:}2$$

(iii) apparent dip along AC i.e.,

$$\text{due S } 66^\circ \text{ W} = \frac{650}{(600-500)} = 1{:}6.5$$

(c) Plotting of Apparent Dips Obtained for Set I

From point O draw OD, OE and OF due N 50°·W· WEST and S 66°W, respectively, which in turn correspond to apparent dips along CB, AB and AC. Mark off apparent dip amounts of 3, 2 and 6.5 along OD, OE and OF, respectively, and obtain points D', E' and F'. Join points D', E' and F' by a straight line which gives the direction of true strike line. Draw a perpendicular OG on the strike line D', E' F', cutting the line at G'. Distance OG' gives the amount of true dip. while OG gives the direction of true dip. It is found to be due N 41° W and the amount of true dip to be 1 : 1.95— (Fig. 2.1.2).

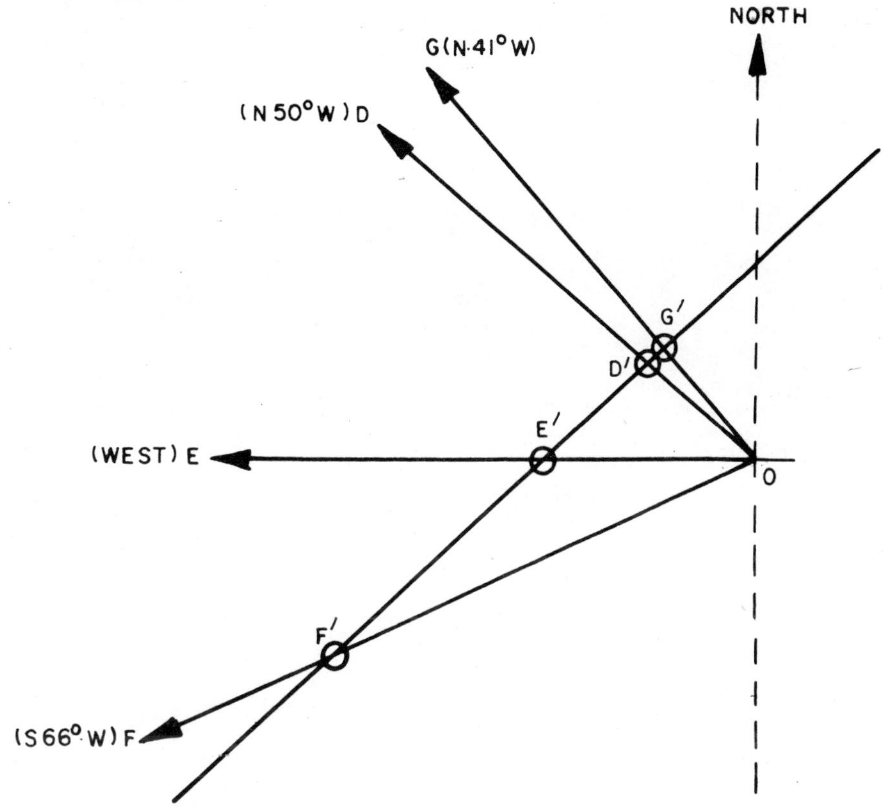

Fig. 2.1.2

Fig. 2.1.2. Plotting Apparent Dips Calculated from Bore Holes *A, B,* and *C,* (Problem 2.1.0)

OD = N 50 °W, OE = West, OF = S 66 °W

OD'= 2 units, OE'= 3 units, OF'= 6.6 units

OG = true dip direction, N 41 °W

$D'E'F'$= true strike direction, N 48 °E - S 48 °W

OG'= true dip amount = 1.95 units

Scale 1 cm = 1 unit

SET II (a) Plotting of boreholes *P*, *Q*, and *R* to a scale of 1 cm = 100 meters, and collection of data regarding distances, bearing between boreholes (Fig. 2.1.3)

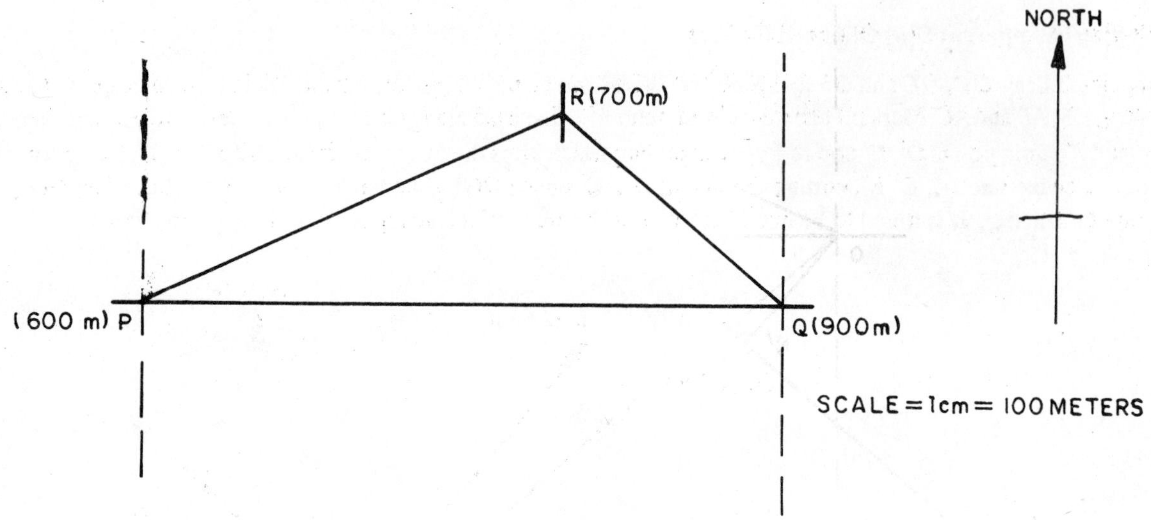

Fig. 2.1.3

Fig. 2.1.3. Plotting of Positions of Bore Holes *P, Q, R*, to A Scale of 1 cm = 100 meters (Problem 2.1.0)

NOTE : In such cases the bearing and the distance between boreholes *P* and *R* is to be measured from the plotting of the data.

(b) Calculation of Apparent Dips for Set II

(i) apparent dip along *PQ* i.e.,

$$\text{due east} = \frac{900}{(900-600)} = 1{:}3$$

(ii) apparent dip along *RQ* i.e.,

$$\text{due S } 50^\circ\text{E} = \frac{400}{(900-700)} = 1{:}2$$

(iii) apparent dip along *PR* i.e.,

$$\text{due N } 67^\circ\text{E} = \frac{650}{(700-600)} = 1{:}6.5$$

(c) Plotting of Apparent Dips Obtained for Set II

From point *O* draw *OK*, *OL* and *OM* due N 67 °E, east and S 50 °E respectively, corresponding in turn to apparent dip along *PR*, *PQ* and *RQ*. Mark off apparent dip amounts of 6.5, 3 and 2 along *OK*, *OL* and *OM*, respectively, and obtain points *K′ L′* and M′. Join the points *K′*, *'L' and M′* by a straight line which gives the direction of true strike line. Draw a perpendicular *ON* and *K′L′M′* from *O*, and cutting the strike line at *N′*. Line *ON* gives the direction of true dip, and the distance *ON′* gives amount of true dip. The true dip is found to be 1 : 1.95, the direction of true dip being due S 41 °E (Fig. 2.1.4).

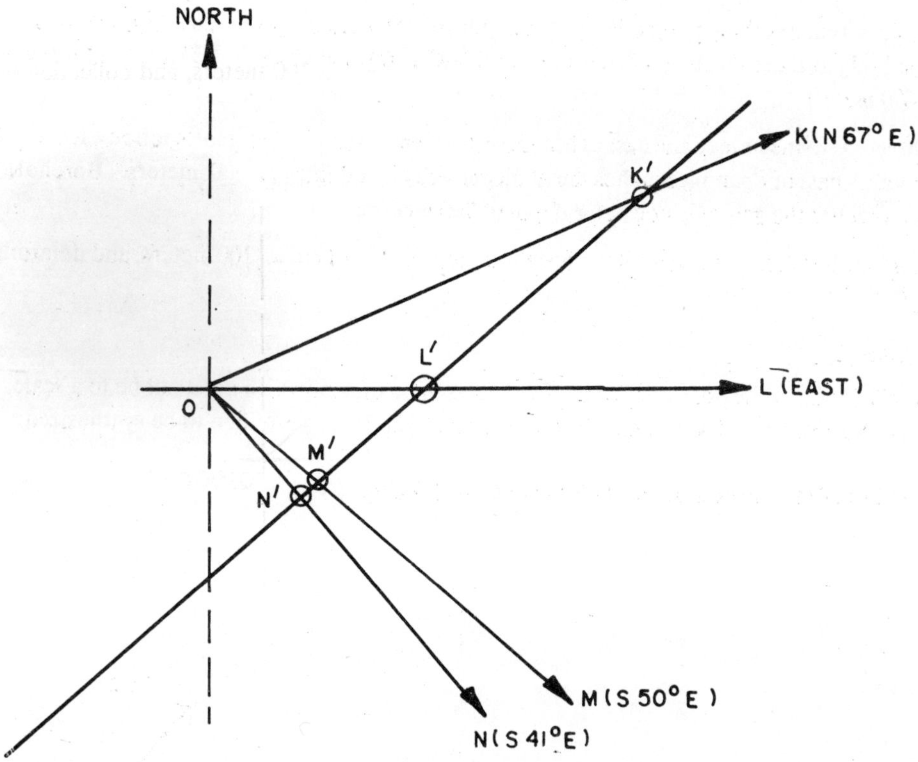

Fig. 2.1.4.

Fig. 2.1.4. Plotting of Apparent Dips Calculated from Bore Holes *P, Q* and *R* (Problem 2.1)

OK = N 67 °E, *OL* = East, *OM* = S 50 °E

OK′ = 6.5 units, *OL′* = 3 units, *OM′*, = 2 units

ON = true dip direction, S 41 °E

K′L′M′= true strike direction N 48 °E – S 48 °W

ON′= true dip amount = 1.95 units

Scale 1 cm = 1 unit

STEP III Synthesis of Results

The results obtained in Step II above, are to be synthesized. Data from Set I yields a true dip direction due N 40 °W, the amount being 1 : 1.95. Data from Set II yields a true dip direction due S 41 °E, the amount of dip being 1 : 1.95. Thus the geologic body possesses dip in exactly opposite directions leading to the inference that a fold exists, it further being an anticline. As the amounts of true dip are the same, the limbs therefore have identical dips. The anticline therefore can be further classified as "symmetrical". The geological structure is hence a "symmetrical anticline". The axis of the anticline trends in N 50 °E - S 50 °W direction.

B. Synclinal Fold

Problem 2.1.1 Two sets of boreholes are sunk to ascertain the structure of a geologic body. Set I is on the western side, and Set II is on the eastern side of an arbitrarily chosen N - S trending line in the field.

SET I Borehole *A* reaches the geologic body at a depth of 800 meters. Borehole *B* is 600 meters west of *A*, and it reaches the geologic body at a depth of 400 meters. Borehole C is 600 meters S 50 °E of B, and it reaches the geologic body at a depth of 600 meters.

SET II Borehole *P* is to the immediate east of borehole A. Here it reaches the geologic body at a depth of 900 meters. Borehole *Q* is 600 meters east of *P*, and it reaches the geologic body at a depth of 500 meters. Borehole *R* is 600 meters N 50 °W of *Q*, and it reaches the geologic body at a depth of 700 meters.

The ground surface is flat. Draw a plan of the boreholes to a scale of 1 cm = 100 meters, and determine the structure of the geologic body.

Procedure

A general plan of the location of the boreholes of the two sets is to be drawn, which need not be to a scale. The attitude of the geologic body is to be determined separately for the two sets. The results are then to be synthesized.

Step I Drawing the plan of the two sets of boreholes (not to scale) (Fig. 2.1.5)

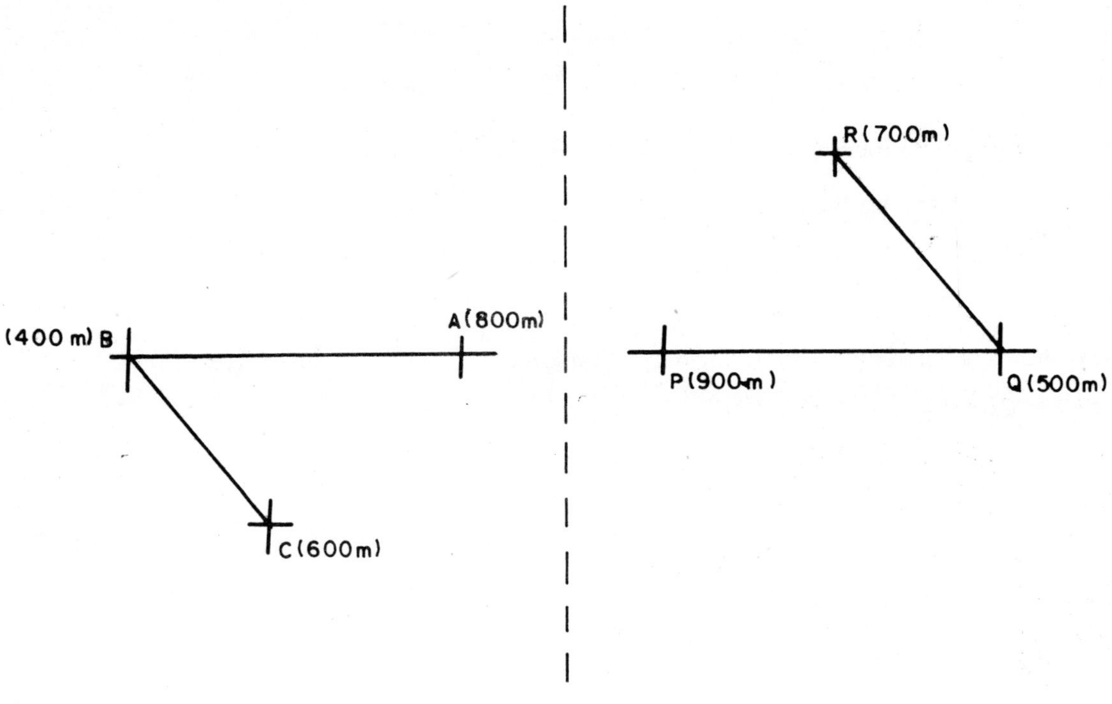

Fig. 2.1.5

Fig. 2.1.5. Plan of Bore Holes *A, B, C* and *P, Q, R* (Not to Scale) (Problem 2.1.1)

NOTE : This helps to plot the data accurately

Step II Determining the Direction and Amount of Apparent Dips, and Deriving the Amount and Direction of True Dip for the two Sets.

SET I (a) Plotting of boreholes *A, B* and *C* to a scale of 1 cm = 100 meters, and collection of data regarding distances, bearing between boreholes (Fig. 2.1.6)

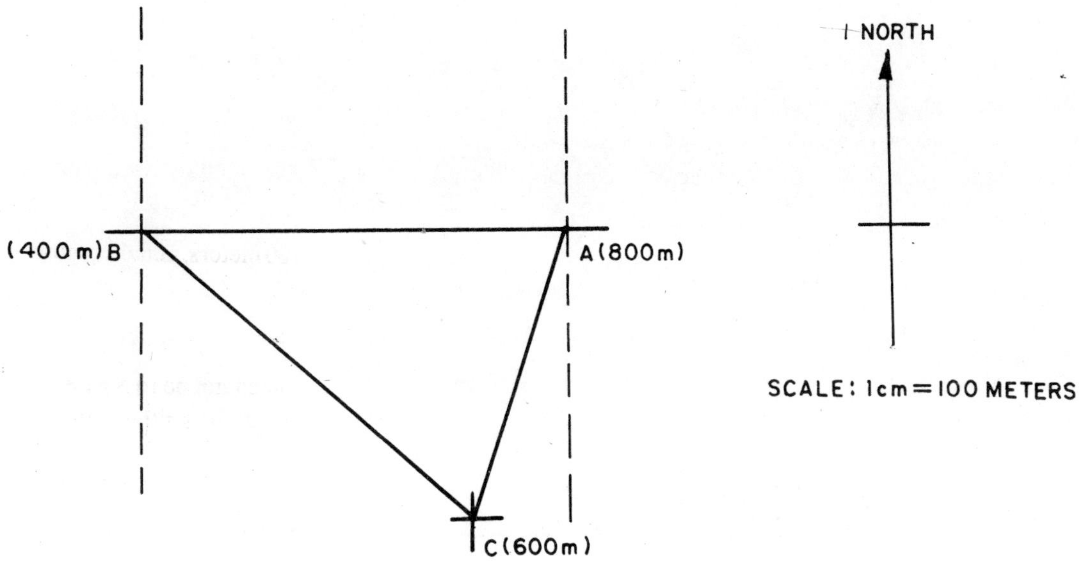

Fig. 2.1.6

Fig. 2.1.6. Plotting of Positions of Bore Holes *A, B, C* to a Scale of 1 cm = 100 meters (Problem 2.1.1)

NOTE : In such cases, the bearing and the distance between borehole *A* and *C* is to be measured from the plotting of the data.

(b) Calculation of Apparent Dips for SET I

(i) apparent dip along *BA* i.e.,

$$\text{due east} = \frac{600}{(800-400)} = 1:1.5$$

(ii) apparent dip along *BC* i.e.,

$$\text{due S } 50^{\circ}\text{E} = \frac{600}{(600-400)} = 1:3$$

(iii) apparent dip along *CA* i.e.,

$$\text{due N } 21^{\circ}\text{E} = \frac{410}{(800-600)} = 1:2.05$$

(c) Plotting of Apparent Dips Obtained for Set I

From point O draw OD, OE and OF due N 21 °E, East and S 50 °E respectively, corresponding in turn to CA, BA and BC, respectively. Mark off the apparent dip amounts of 2.05, 1.5, and 3 along OD, OE and OF respectively, and obtain points D' , E', and F'. Join the points D', E and F'' by a straight line, which gives the direction of true strike line. Draw OG perpendicular to $D'E'F'$, which gives the direction of true dip. OG cuts the strike line at G' and OG' gives the amount of true dip. It is found to be 1 : 1.4 and the direction of true dip to be due N 67 °E (Fig. 2.1.7)

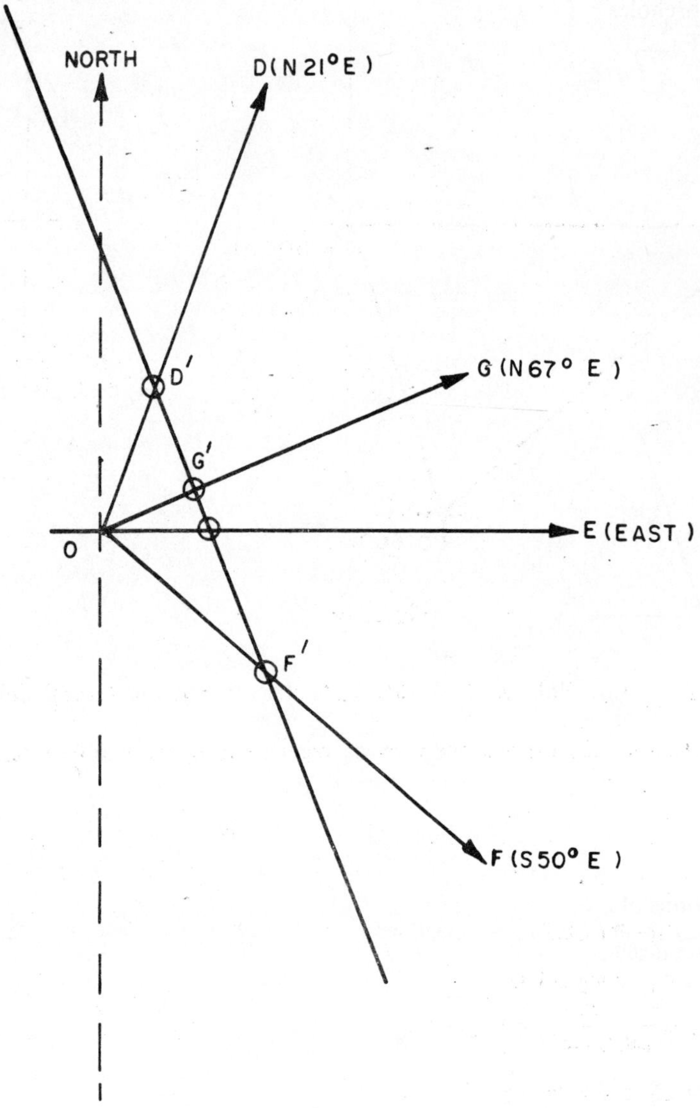

Fig. 2.1.7

Fig. 2.1.7. Plotting of Apparent Dips Calculated from Bore Holes *A, B* and *C* (Problem 2.1.1)

OD = N 21 °E, OE = East, OF = S 50 °E

OD' = 2.05 units, OE' = 1.5 units, OF' = 3 units

OG = true dip direction, N 67 °E

$D'E'F'$ = true strike direction, N 23 °W - S 23 °E

OG' = true dip amount = 1.4 units

Scale 1 cm. = 1 unit

SET II (a) Plotting of boreholes *P*, *Q* and *R* to a scale of 1 cm = 100 meters and collection of data regarding distances, bearing between boreholes (Fig. 2.1.8).

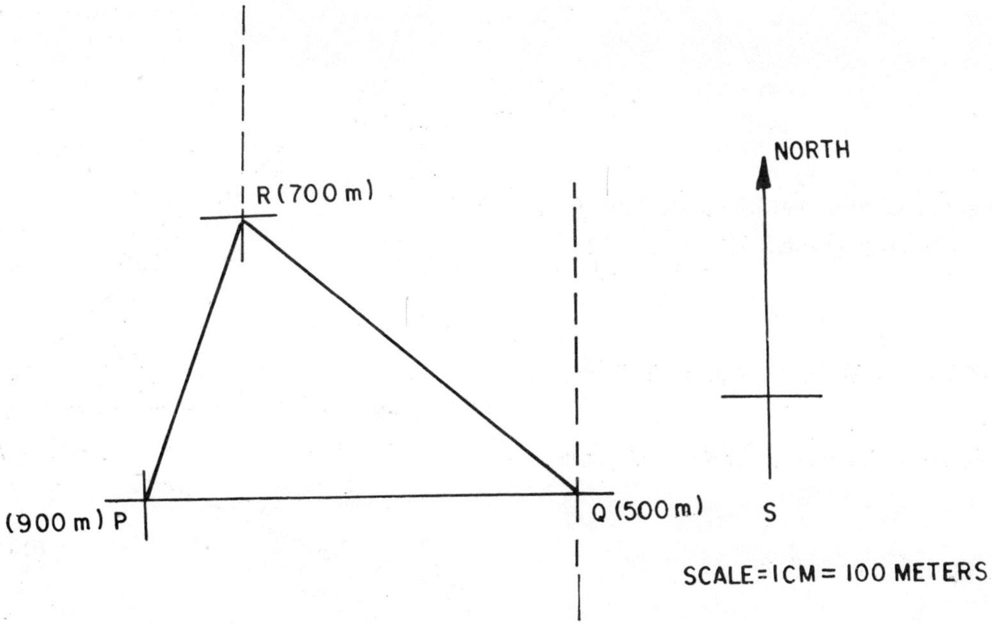

Fig. 2.1.8

Fig. 2.1.8. Plotting of Positions of Bore Holes *P, Q, R* to a Scale of 1 cm = 100 meters (Problem 2.1.1)

NOTE : In such cases, the distance and the bearing between boreholes *P* and *R* is to be measured from plotting of the data.

(b) Calculation of apparent dips for Set II

(i) apparent dip along *QP*

i.e., due west $= \dfrac{600}{(900-500)} = 1:1.5$

(ii) apparent dip along *QR*

i.e., due N20 °W $= \dfrac{600}{(700-500)} = 1:3$

(iii) apparent dip along *RP*

i.e., due S 20 °W $= \dfrac{410}{(900-700)} = 1:2.05$

Plotting of aparent dips obtained for Set II

From point *O* draw *OK*, *OL* and *OM* due N 50 °W, west, and S 20 °W respectively, corresponding in turn to *QR*, *QP* and

RP respectively. Mark off apparent dip amounts 3, 1.5, and 2.05 along *OK*, *OL* and *OM* respectively, and obtain points K′, *L′* and *M′*. Join these points by a straight line which gives the direction of true strike line. Draw *ON* perpendicular from *O* on *K′L′M′* which gives the direction of true dip. The line *ON* cuts *K′L′M′* at *N′*. The distance *ON′* gives the amount of true dip. It is found to be 1 : 1.4 and the direction of true dip to be S 68 °W (Fig. 2.1.9).

Fig. 2.1.9. Plotting of Apparent Dips Calculated from Bore Holes *P*, *Q* and *R* (Problem 2.1.1)

OK = N 50 °W, *OL* = West, *OM* = S 20 °W

OK′ = 3 units, *OL′* = 1.5 units, *OM′* = 2.05 units

ON = true dip direction, S 68 °W

K′L′M′ = true strike direction, N 22 °W – S 22 °E

ON′ = true dip amount = 1.4 units

Scale 1 cm = 1 unit

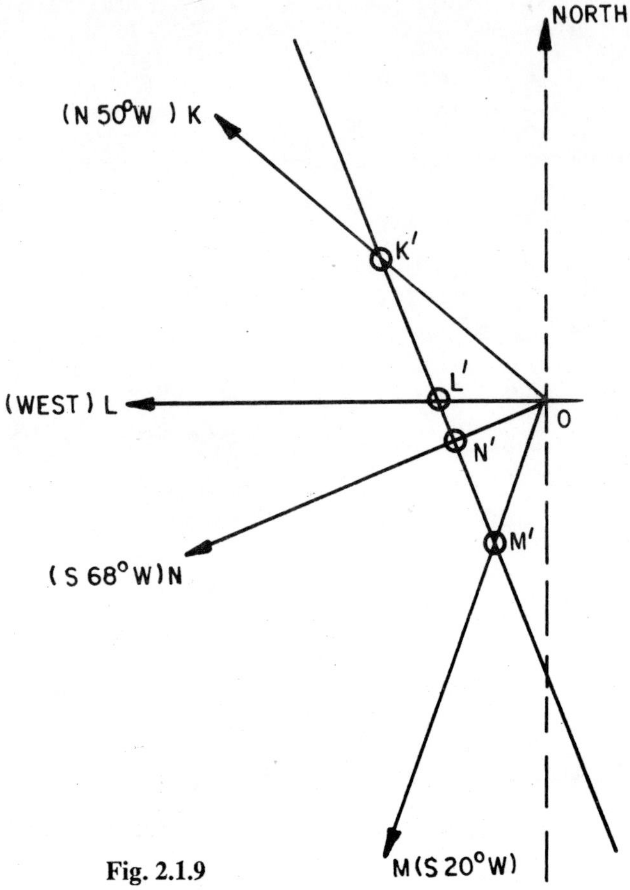

Fig. 2.1.9

Step III Synthesis of Results

Results from Steps I, II show that the geologic body has a dip of 1 : 1.4 due N 67 °E for the western block boreholes, while it has a dip of 1 : 1.4 due S 68 °W for the eastern block boreholes. Thus the two directions of the dip point towards a common point i.e., the two limbs of the fold are dipping towards each other. Hence it represents a synclinal structure. Further as the amounts of dip are same for the two limbs, the fold belongs to the category of a symmetrical type. Further the axis of the fold trends in N 22 °W - S 22 °E direction.

2.1.2 OTHER TYPE OF STRUCTURES

In the above described structures (Problem 2.1.0 and 2.1.1.) the presence of a fold was clearly indicated because the two sets of boreholes were separated from one another. In actuality, in the field when boreholes are sunk, one does not know about the existence or otherwise of a fold. As such the borehole data has to be handled carefully and with good deal of imagination.

Several modes of representation of structures have been presented in Figs. 2.2. 0 *A*, *B*, *C*, *D*, 2.2.1 *A*, *B*, *C*, *D*, 2.2.2., 2.2.3, 2.2.4 and 2.2.5. In some representations, the direction of increase in the amount of dip has been noted (Fig 2.2.0 *A*, *B*, *C* and *D*). In others, length of the arrows indicates the increase in the amount of dip (Fig. 2.2.1 *A*, *B*, *C*, and *D*, 2.2.2 and 2.2.3).

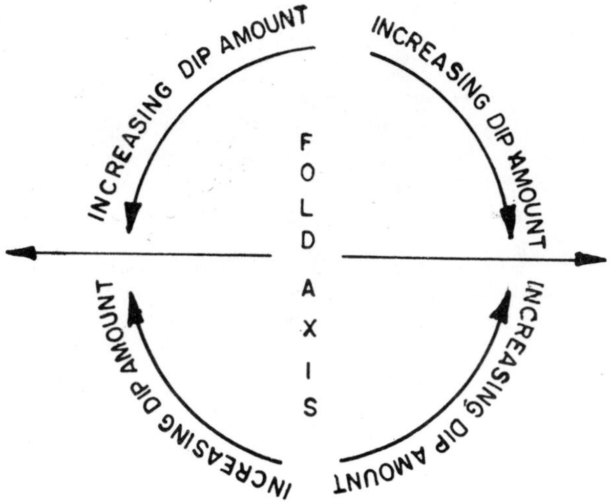

Fig. 2.2.0 (A)

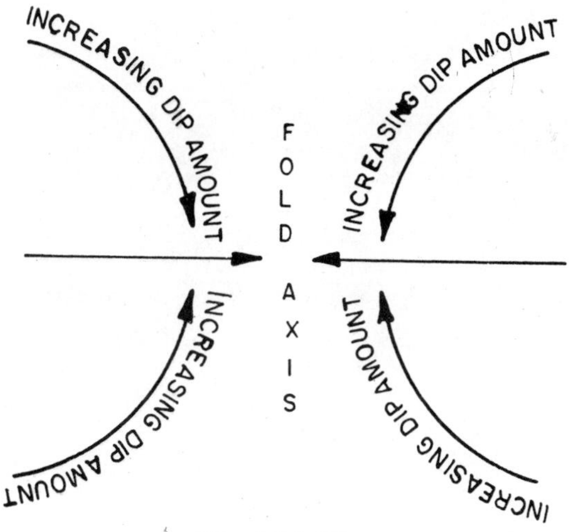

Fig. 2.2.0 (B)

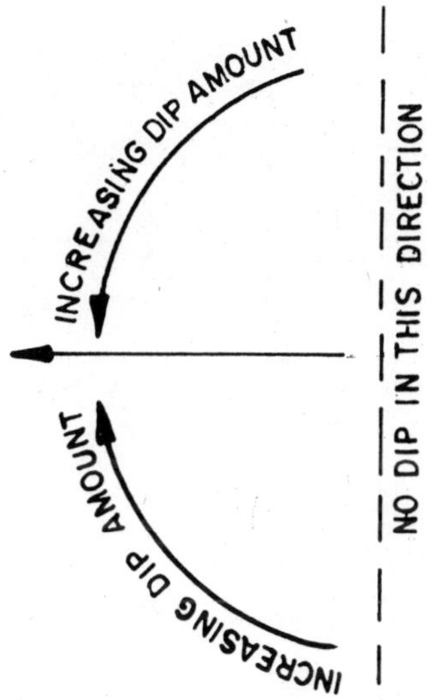

Fig. 2.2.0 (C)

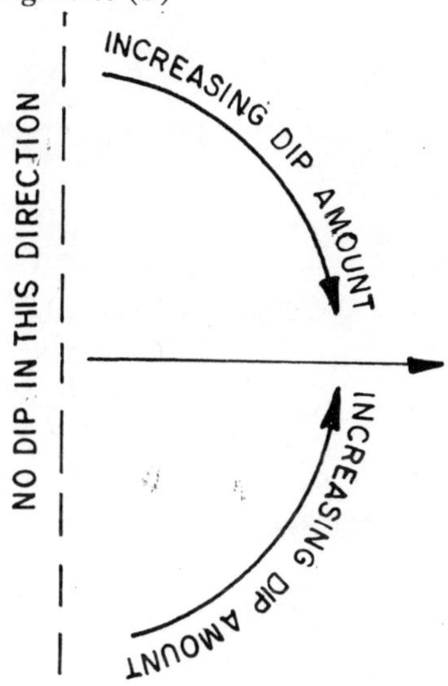

Fig. 2.2.0 (D)

Fig. 2.2.0 (A)

Disposition of dip amount in an anticlinal fold. Note that maximum dip is due west for the western limb, and maximum dip is due east for the eastern limb of the fold. The fold axis is therefore trending N–S and in this direction, there is no dip for the two limbs.

Fig. 2.2.0 (B)

Disposition of dip amount in a synclinal fold. Note that the maximum dip is due east for the western limb of the fold, and it is due west for the eastern limb of the fold. In this case also the axis of the fold is trending N-S.

Fig. 2.2.0 (C)

Diposition of dip amount in a singularly dipping formation. Note that the dip is maximum due west. Also there is no dip due east, at all.

Fig. 2.2.0 (D)

Disposition of dip amount in a singularly dipping formation. Note that the dip is maximum due east. Also there is no dip due west, at all.

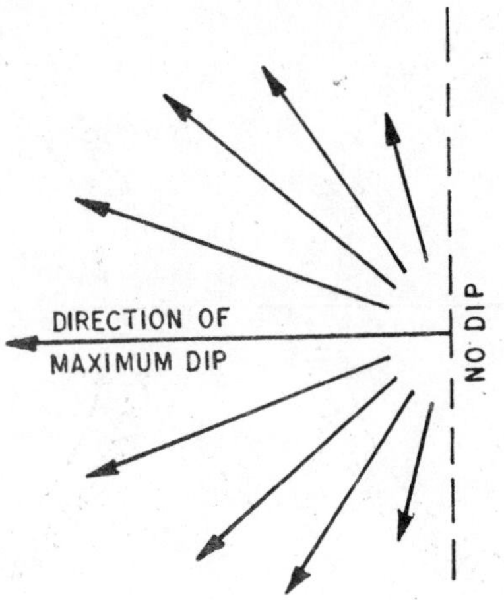

Fig. 2.2.1 (A)

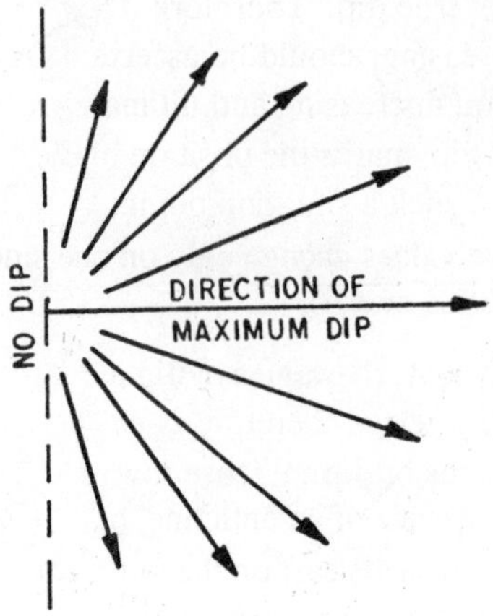

Fig. 2.2.1 (B)

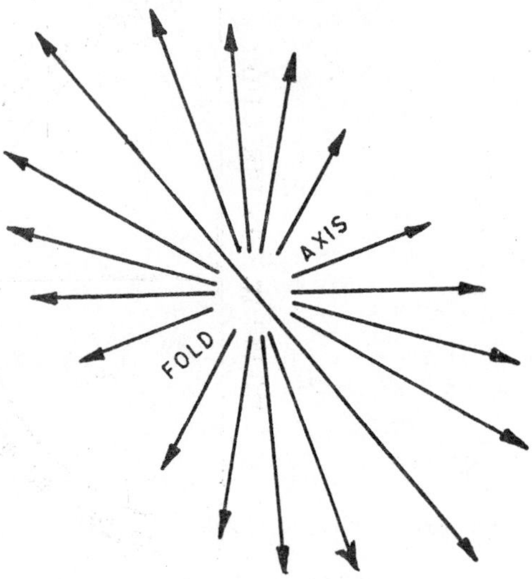

Fig. 2.2.1 (C)

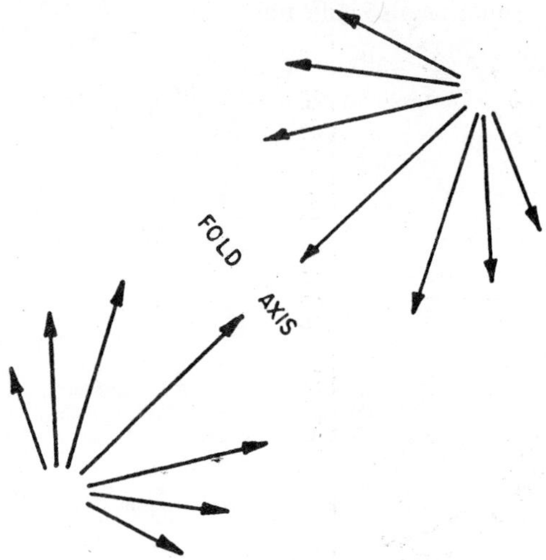

Fig. 2.2.1 (D)

Fig. 2.2.1 (A)
Diposition of dip amount in a singularly dipping formation. Note that the increase in the amount of dip is indicated by the increased length of the arrow. Maximum dip is due west. Also there is no dip due east, at all.

Fig. 2.2.1(B) Diposition of dip amount in a singularly dipping formation. Note that the increase in the amount of dip is indicated by the increased length of the arrow. Maximum dip is due east. Also there is no dip due west, at all.

Fig. 2.2.1 (C) Disposition of dip amount in an anticlinal fold. Note that the length of the arrow indicates the amount of dip The length of the arrow is maximum towards northwestern direction (for the western limb) and towards southeastern direction (for the eastern limb). The fold therefore is an anticline with a NE–SW trending axis. Note further that along axis, the fold is not having any inclination. It is a non-plunging fold.

Fig. 2.2.1 (D) Deposition of dip amount in a synclinal fold. Note that the length of the arrow indicates amount of dip. The length of the arrow is maximum towards northeastern direction (for the western limb), and it is maximum towards southwestern direction (for the eastern limb). The fold therefore is a syncline with a NW-SE trending axis. Note further that along the axis, the fold is not having any inclination. It is a non-plunging fold.

A study of the disposition of the values of apparent dip (Fig. 2.2.0 *A*, *B*, *C* and *D*) reveals that these increase towards the direction of true dip. Therefore if several borehole data be available, then the positions of the directions in which the values are increasing, should be ascertained. It is to be remembered that on the two sides of the position of the true dip, the values go on decreasing and ultimately the value will be zero (Fig. 2.2.0 *C* and *D*, Fig. 2.2.1 *A* and *B*). The position of zero apparent dip, marks the position of the true strike line. If across such a position, the values of the apparent dips were to change, then such a situation points to the presence of fold - either anticline or syncline. (Fig. 2.2.0 *A* and *B*, Fig. 2.2.1 *C* and *D*). If the values change only on one side, then it indicates the presence of singularly dipping formation (Fig. 2.2.0 *C* and *D*, Fig. 2.2.1 *A* and *B*).

In Fig. 2.2.0 *A*, the values of dip increase towards west (left hand side) as well as towards east (right hand side). The structure to be inferred therefore is an anticline with its axis trending in N - S direction (Fig. 2.2.0 *A*). Likewise in Fig. 2.2.1 *C*, the values of dip increase towards northwest side, and also towards southeast side. Therefore this situation also indicates the presence of an anticline, but its axis is trending in NE - SW direction (Fig. 2.2.1 *C*). In Fig. 2.2.0 *B*, the values of dip increase towards east on the left hand side of the fold, while these are observed to increase towards west on the right hand side of the fold. The structure to be inferred from such a situation is a syncline with a N - S trending axis (Fig. 2.2.0 *B*). In Fig. 2.2.1 *D* the dip is found to increase due northeast for the western limb, and due southwest for the eastern limb. Therefore this situation also indicates the presence of a syncline with its axis trending in NW - SE direction (Fig. 2.2.1 *D*).

Structures are generally complex. Folds may plunge in one direction or in two directioins. The folds may be further refolded. Sructural Domes and basins also may exist. All these complications will affect the values of dip and their spatial dispositions. Therefore from the borehole data, derivation or the inference in respect of what type of structure is actually

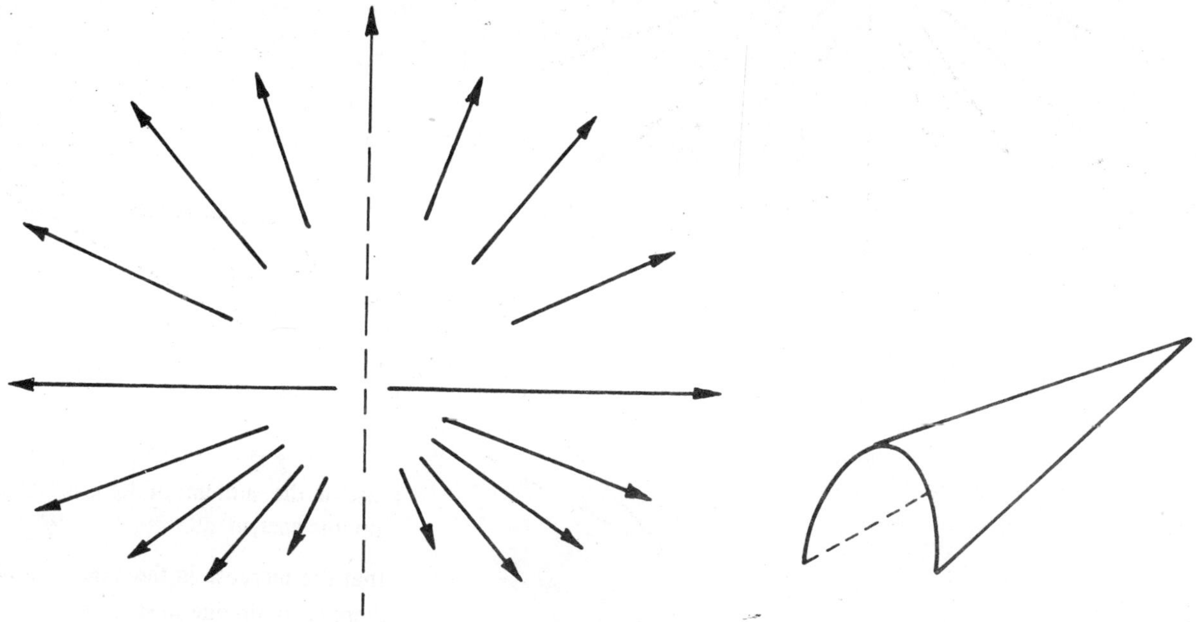

Fig. 2.2.2

Fig. 2.2.2 Disposition of dip amount in a singularly plunging anticlinal fold. Note that though the amount of maximum dip increases towards west (for the western limb) and towards east (for the eastern limb), the fold has some dip towards north, but none towards south. This direction actually marks the direction of strike, and there should not have been any inclination in this direction. Therefore even if the dip decreases away from the direction of full dip (true dip direction), the limbs still possess some inclination in the direction of axis of the fold (in the present case the axis has inclination towards north). The structure therefore is a singularly plunging anticlinal fold.

present, requires a very careful study of the values of dip and their directions of dispositions. The geoscientists also should be aware of the implications of the different structures upon the values of dip. Numerous boreholes are therefore necessary. Ultimately it is a matter of experience as far as the establishment of structure is concerned. But the approach should be scientific, methodic and logical one. Simplified behaviour of the dips in respect of singularly plunging anticline, doubly plunging eanticline, structural dome and structural basin are schematically shown in Figs. 2.2.2, 2.2.3, 2.2.4 and 2.2.5, respectively.

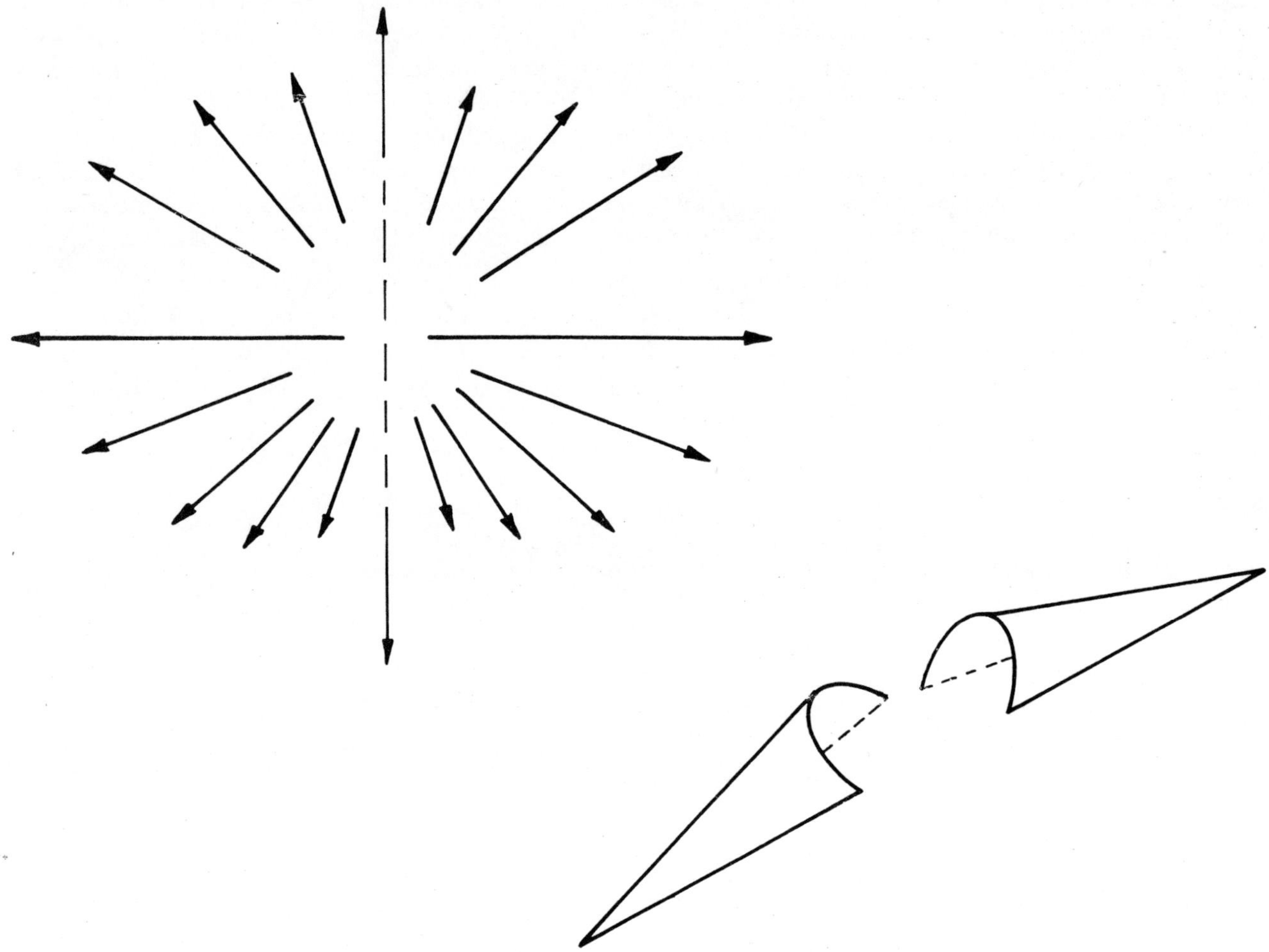

Fig. 2.2.3

Fig. 2.2.3 Disposition of dip amount in a doubly plunging anticlinal fold. Note that though the amount of maximum dip increases towards west (for the western limb), and towards east (for the eastern limb), the axis of the fold has some inclination towards north, as well as towards south. This direction actually marks the direction of strike, and there should not have been any inclination in this direction. Therefore, even if the dip decreases away from the direction of full dip (true dip direction), the limbs still possess some inclination due north, as well as due south. The structure therefore is a doubly plunging anticlinal fold.

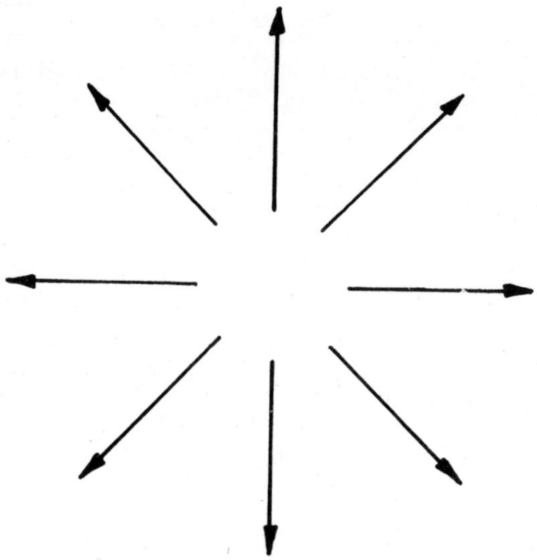

Fig. 2.2.4

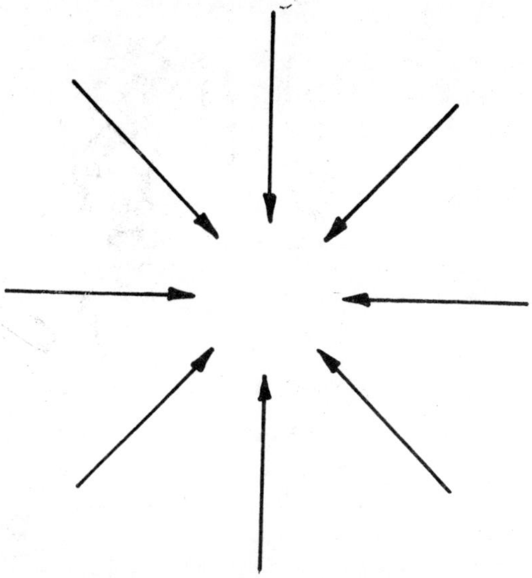

Fig. 2.2.5

Fig. 2.2.4 Disposition of amount of dip in a structural dome. Note that the formation has indentical inclination (length of each arrow is same) in all directions, which is not possible either in a singularly dipping formation, or a folded formation. The dip is away from a central point.

Fig. 2.2.5 Disposition of amount of dip in a structural basin. Note that the formation has indentical inclination (length of each arrow is same) in all directions, which is not possible either in a singularly dipping formation, or a folded formation. The dip is towards a common central point.

3

ESTIMATION OF THICKNESS, DISTANCE AND DEPTH OF OREBODY, GEOLOGIC BODY ETC.

3.1.0 INTRODUCTION

The geologist is called upon to estimate the thickness of formation, or the depth at which the top of a formation can be encountered. This can be achieved only through the knowledge of true dip, its direction and the nature of ground surface. In the following pages the procedure to be adopted in measuring the thickness, distance, depth etc., will be elaborated.

3.1.1 Thickness

This is required to be estimated in the case of bedded formation like the sedimentary rocks, a sill, lava flow or a dyke. The thickness is the perpendicular distance between the top and bottom of a bed. This measurement also is needed to be made only in the true dip direction, if the formation be dipping one. The exposure of a bed on a plane or horizontal ground surface, and under different inclinations of a bed is shown in Fig. 3.1

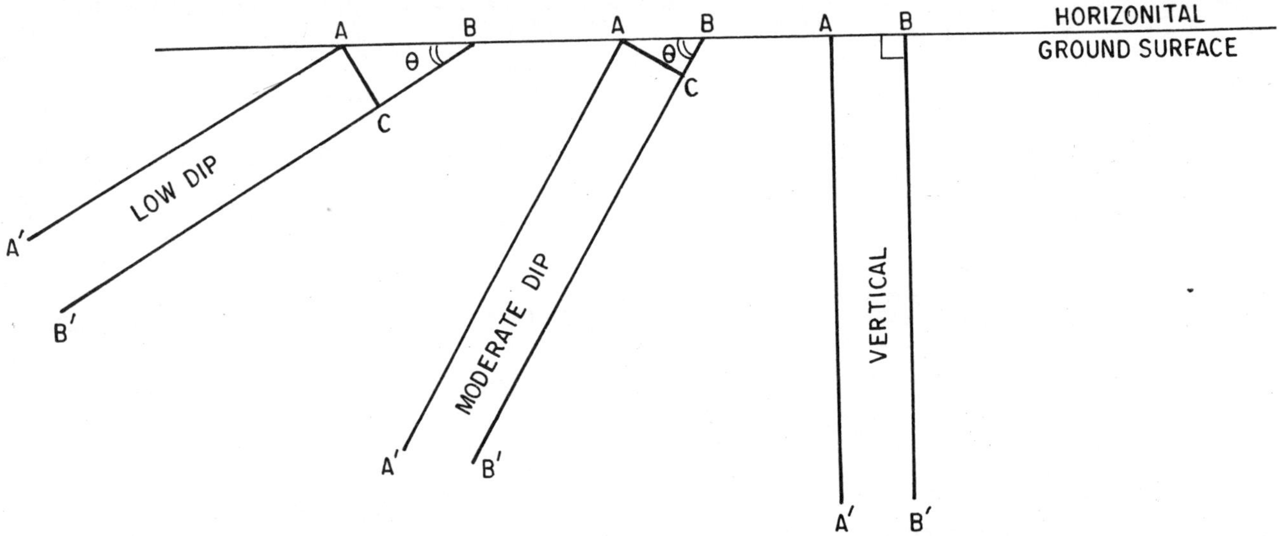

Fig. 3.1

Fig. 3.1 CONTROL OF DIP AMOUNT OF THE WIDTH OF OUTCROP

The formation is one centimeter thick in all the drawings, but the width of the outcrop (*AB*) is not so. It is 2.15 cm, 1.12 cm and one cm corresponding to low dip, moderate dip and vertical dip, respectively. *AB* is the width of the outcrop, whereas *AC* is the thickness.

In Fig. 3.1 the thickness *AC* is same, but the inclination of the bed is different. The width of the outcrop *(AB)* is also different - maximum with low dip, intermediate with moderate dip and least, and which is equal to the thickness with vertical dip (Fig. 3.1). When the true dip is known and the ground surface is flat, then a simple mathematical relation exists between the width of the outcrop and the thickness. This has been shown below.

Thickness (of a bed) = Sin Ø (inclination of bed) x width of the outcrop

Therefore by referring to the mathematical tables, the thickness can be estimated by only measuring the actual width of the outcrop in the field. Even the value of Sin (*theta* is the angel of dip of a bed) need not be found out. One can plot the width of the outcrop (line *AB*) to a scale, and then construct the angle of dip (θ) at *B*, and at *A*. Afterwards draw lines *AA′* and *BB′* (Fig. 3.1). Draw a perpendicular from *A* on *BB′* and measure the distance *AC* (thickness of the bed) to the same scale as that taken for the distance *AB*. In case the body is vertical, then the width of the outcrop itself stands for the thickness (Fig. 3.1).

3.1.2 Estimation of Depth and Horizontal Distance

Problem 3.1.0 In borehole *A*, a formation is met with at a depth of 600 meters. The formation has a dip of 1:4 due N 40°W, Estimate the depth of formation in borehole *B* which is located 500 meters east of borehole *A*. At what distance south of borehole *A*, formation will be encountered at the surface? Also estimate the depth of the formation in borehole *D* which is located 1000 meters S 80 °W of borehole *A*. The ground surface is flat.

Procedure

It is necessary to draw a rough sketch (not to scale) of the position of the different boreholes in order to adopt the further steps. Since the true dip has been already given, it is possible to easily estimate the amount of dip in any other discretion. The other details are described further.

Step I **Drawing of rough sketch of the positions of different boreholes (Fig. 3.2)**

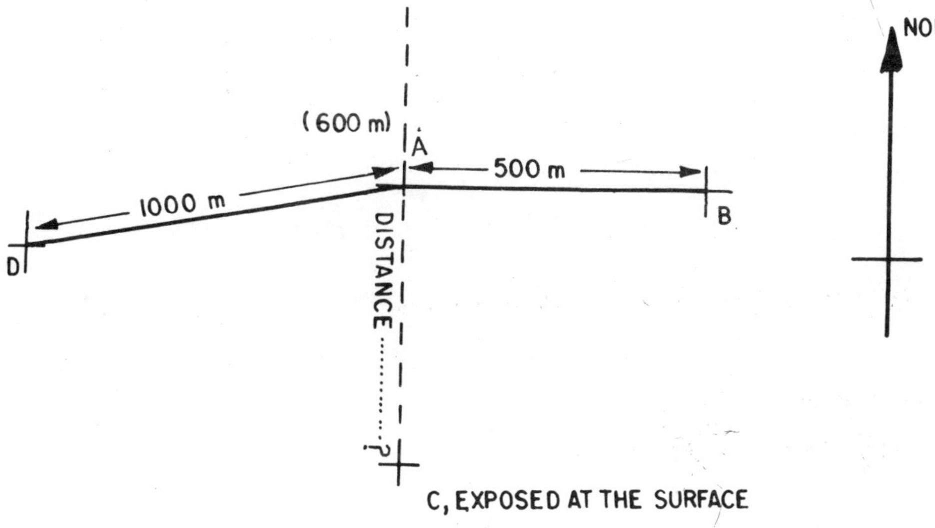

Fig. 3.2

Fig. 3.2 SKETCH OF POSITIONS OF *A*, *B*, *C* AND *D* (NOT TO SCALE)

(Problem 3.1)

Step II Plotting of True Dip and Finding out the Amount Apparent Dips

From point O draw OP due N 40 °W and scale out 4 units along it and obtain point P' (Fig. 3.3). Through P' draw a perpendicular on OP, this line giving the direction of true strike. Draw OQ due east (because it is necessary to estimate the amount of apparent dip in that direction, since borehole B is east of borehole A) but such a line cannot cut the strike line. Therefore extend it in opposite direction i.e., due west, till it cuts the strike line at Q'. Direction OQ' i.e., east is not

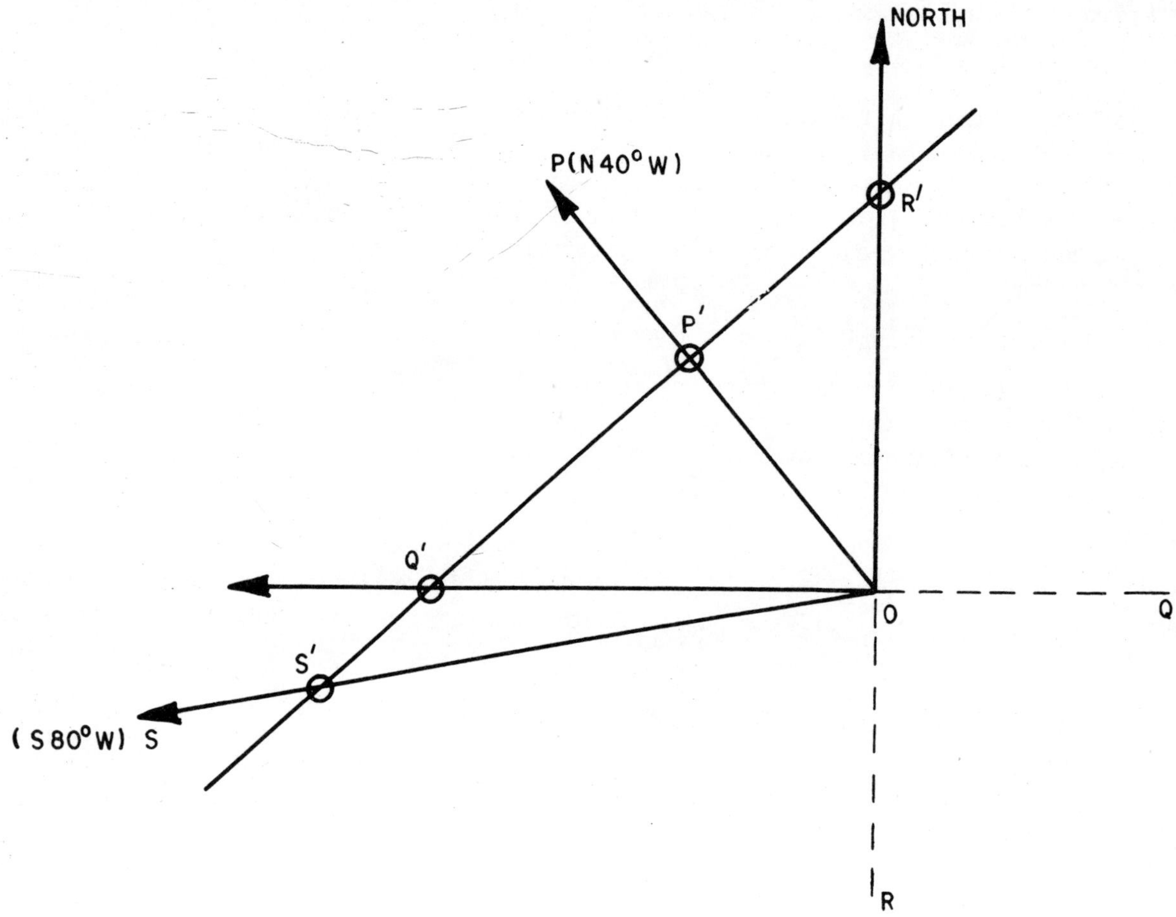

Fig. 3.3

Fig. 3.3 PLOTTING OF TRUE DIP OF 1 : 4 DUE N 40 °W (Problem 3.1)

OP = N 40 °W OP' = 4 units

OR = apparent dip due South

OR' = anti dip amount along OR = 5.3 units

OQ = apparent dip due east

OQ' = anti dip amount along OQ = 6.15 units

OS = apparent dip due S 80 °W

OS′ = apparent dip amount = 7.8 units

Scale 1 cm = 1 unit

apparent dip, but it is "anti apparent dip". Likewise draw OR due south (because it is necessary to ascertain the amount of apparent dip in that direction, since point C is south of borehole A, as per data), but this line also cannot cut the strike line. Therefore extend it towards north (the line OR' in Fig. 3.3). There is therefore no apparent dip due south, but it marks the "anti apparent dip" direction. Measure the anti apparent dip amounts. These are found to be 1 : 6.15 and 1 : 5.3 due west and north respectively (OQ' and OR' in Fig. 3.3). Draw line OS due S 80 °W (this direction marks the position of borehole D with respect to borehole A, Fig. 3.3) which cuts the strike line at S'. OS' is found to be 7.8 units long, and this line directly cuts the strike line, unlike the lines OR and OQ. Therefore the direction S 80 °W marks apparent and not "anti apparent dip direction".

Step III Estimation of Depths in Boreholes B and D

Borehole B This can be ascertained by considering the depth in borehole A and the amount of apparent dip along AB. B is east of A and therefore the amount of apparent dip due east was tried in Step II above by drawing the line OQ in Fig. 3.3. However it has been found that in the easterly direction there is no dip, but it is only "anti dip". This means that in the easterly direction, the formation will be encountered at a level shallower than that encountered in borehole A. The rate of shallowing is 1 : 6.15 (amount of apparent dip along OQ' in Fig. 3.3). The distance between A and B borehole is given as 500 meters. The amount of raise over this distance is calculated by the below noted equation.

$$\frac{\text{Distance between boreholes } A \,\&\, B \text{ x 1}}{\text{6.15 (rate of apparent dip)}} = \frac{500 \text{ x } 1}{6.15} = 8.13 \text{ meters}$$

The raise in the level between the boreholes A and B is of 81.3 meters. But the formation is at a depth of 600 meters in borehole A. Therefore 81.3 is to be subtracted from 600 meters (depth in borehole A). The depth in borehole B will be 518.7 meters (600 minus 81.3).

Borehole D This borehole is S 80 °W of borehole A. Therefore find the amount of apparent dip along S 80 °W, as is described in Step II above. Because S 80 °W happens to be " apparent dip" and not "anti apparent dip" direction, the formation will be encountered in deeper level as the distance increases. The rate of apparent dip is found to be 1 : 7.8 (line OS' in Fig. 3.3.), and the distance between boreholes A and D is given as 1000 meters. The amount of lowering is estimated by using the following equation.

$$\text{Amount of lowering} = \frac{\text{Distance x 1}}{\text{7.8 (rate of apparent dip)}} = \frac{1000 \text{ x } 1}{7.8} = 128.2 \text{ meters}$$

The value of 128.2 meters is with respect to the depth in borehole A. The formation is already at a depth of 600 meters in borehole A. Hence in the borehole D the same formation will be encountered at a still deeper level, as the value of 128.2 meters is to be added, because borehole D is in the "apparent dip direction" and not in the "anti apparent dip" direction with respect to borehole A.

600 (depth in borehole A) + 128.2 (amount of lowering over 100 meters) = 728.2 (meters)

This information can then be utilized to estimate the depths in the other boreholes, or the distance at which a particular depth can be reached, and so on.

3.1.3 Estimation of Distance with Respect to Borehole A and South of it, Where the Formation will Crop out at the Surface

This also needs the apparent dip along southern direction. This has been found out to be 1:5:3 (line OR' in Fig. 3.3). However there is no apparent dip due south, but it is "anti apparent dip". This indicates that the formation will be encountered at shallower level with increased distance due south of borehole A. The formation is required to be at the surface, but this is with respect to borehole A Hence the depth of the formation in that borehole should be known, which is given as 600 meters. Therefore the formation is required to be raised over a level 600 meters. The equation used in such calculations is given below, which involves the apparent dip of 1:5.3 due south.

$$\text{Horizontal distance} = \frac{\text{rate of apparent dip x amount of raise}}{1}$$

$$= \frac{5.3 \times 600}{1} = 3180 \text{ meters}$$

Therefore at a distance of 3180 meters south of borehole *A*, the formation will be exposed at the surface.

From the procedure described above, it is clear that in the estimation of depth, or distance, the value of true dip and also the apparent dips are absolutely necessary. In the example cited above (Problem 3.1), the value of true dip and the direction was given directly. However in actuality with the help of 2 or 3 boreholes it will be first necessary to calculate the direction and the amount of true dip. This information can then be utilized to estimate the depths in the other bore holes, or the distances at which a particular depth can be reached, and so on.

4

DETERMINATION OF THROW AND NATURE OF FAULTS

4.1.0 INTRODUCTION

The rocks are often faulted. This necessiates the determination of the amount of throw and the side of downthrow. Basically this can be determined provided the attitude of the rock is available, or is determinable from the borehole data. Borehole data across the fault line is needed to be given, then alone the throw and the downthrow side can be estimated. When two or more faults be present, then the kind of the faults also can be ascertained. The procedure to achieve all this will be described in the following paragraphs.

4.1.1 Case I Presence of only one Fault

Problem 4.1 In a flat terrane 4 boreholes are sunk to ascertain the existence of a fault. In borehole *A* the formation is reached at a depth of 300 meters. Borehole *B* is 500 meters S 10 °W of *A*, and it reaches the formation at a depth of 500 meters. Borehole *C* is 800 meters east of *B* and here it reaches the formation at a depth of 900 meters. Borehole *D* is located 200 meters N 70 °E of *C* and the same formation is reached at a depth of 200 meters. Between boreholes *C* and *D*, a fault trending N 15 °E - S 15 °W is suspected. Determine the attitude of the dipping formation and also determine the kind of fault, amount of throw and the downthrow side. Draw a plan of the boreholes and of the fault to a scale of 1 cm = 100 meters. The ground surface is flat.

Procedure

Such problems are to be solved in two stages. In the first stage, the attitude of the dipping formation is to be determined, considering only those boreholes that are on one and the same side of the fault plane. In the second stage, with the knowledge of true dip direction, the depth of the formation in a borehole situated across the fault line, is ascertained, considering as if faulting had not taken place. The value so obtained is called the "expected value" and this is compared with the one actually given in the data. This helps to ascertain the amount of throw and the down throw side. The attitude of the dipping formation determined by calculation is compared with the trend of the fault line and the fault is then classified. The details of the procedure are elaborated below.

Step I Determining Attitude of the Formation

(a) Drawing the plan of the boreholes to the scale of 1 cm = 100 meters (Fig. 4.1.0) and determining the distance and bearing between boreholes *A* and *C*.

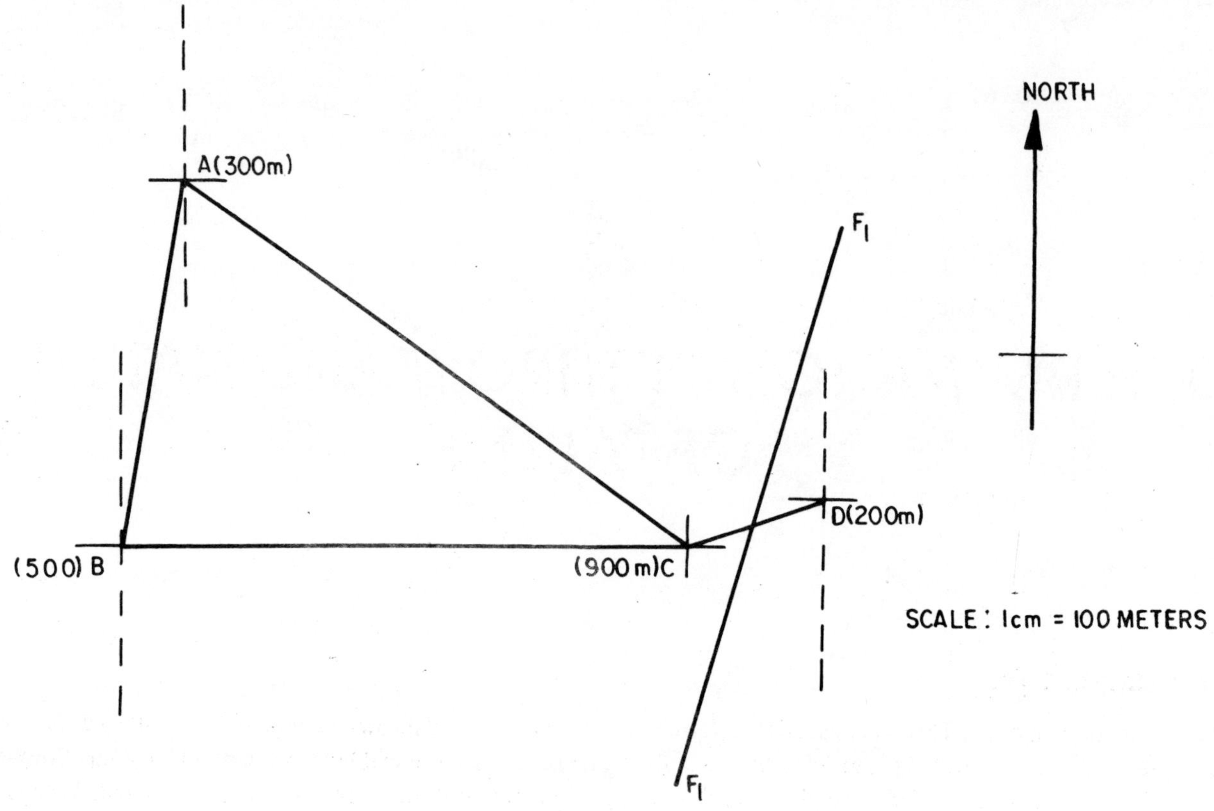

Fig. 4.1.0

Fig. 4.1 PLOTTING OF POSITIONS OF BORE HOLES *A, B, C,* and FAULT F_1 TO A SCALE OF 1 cm = 100 meters (Problem 4.1)

NOTE : In such cases, the distance and the bearing between boreholes *A* and *C* is to be measured from the plotting of the data.

(b) *Calculation of apparent dips considering boreholes A, B, and C*

(i) apparent dip along *AB*

i.e., due S 10°W = $\frac{500}{(500-300)} = 1:2.5$

(ii) apparent dip along *BC*

i.e., due east $\frac{800}{(900-500)} = 1:2$

(iii) apparent dip along *AC*

i.e., due S 55°E = $\frac{870}{(900-300)} = 1:1.45$

(c) Plotting of apparent dips obtained between boreholes A,B,C

From point *O* draw *OP*, *OQ* and *OR* due East, S 55 °E and S 10 °W respectively. Mark off apparent dip amounts of 2, 1.45 and 2.5 on *OP*, *OQ* and *OR*, and obtain points *P′*, *Q′* and *R′*. Join *P′*, *Q′* and *R′* by a straight line. Draw *OS* perpendicular from *O* on *P'Q'R'*, and obtain points *S'*. Measure the dip amount (*OS'*) and direction *OS'S*. It is found to be 1:1.42, the direction of dip being S 45 °E (Fig. 4.1.1).

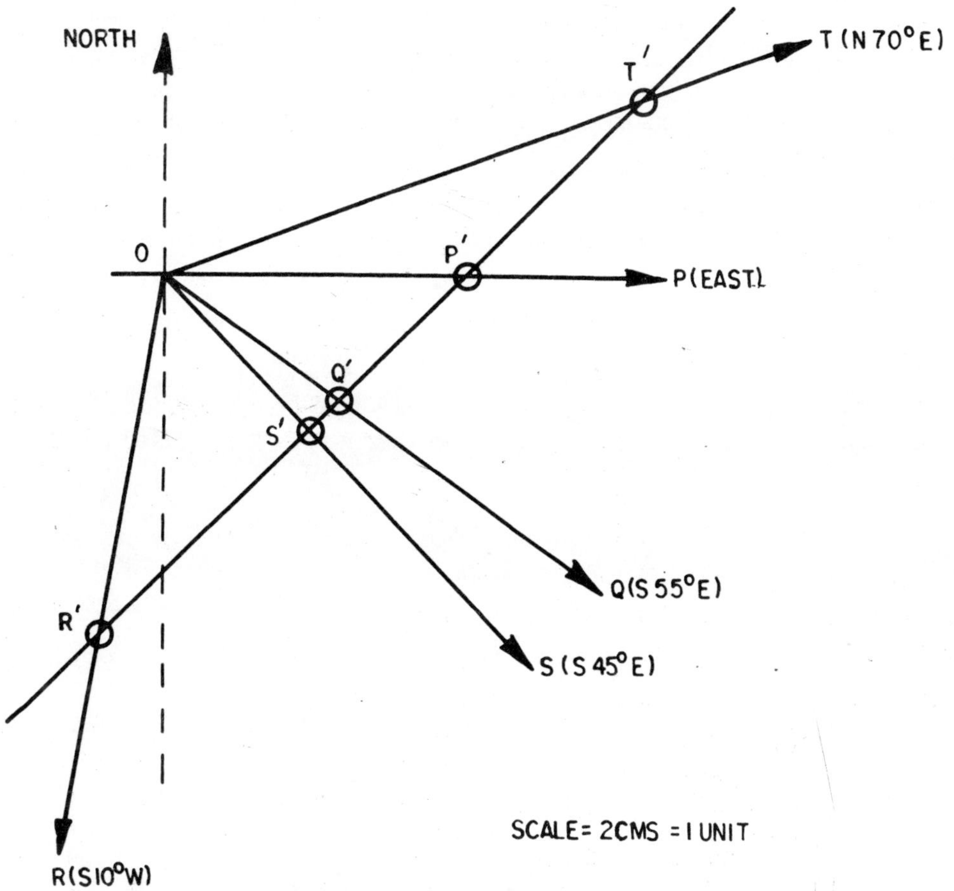

Fig. 4.1.1

Fig. 4.1.1 Plotting of Apparent Dips (Problem 4.1.1)

OP = East, *OQ* = S 55 °E, *OR* = S 10 °W

OP' = 2 unit *OQ'* = 1.45 units, *OR'* = 2.5 units

OS = true dip direction, 545°E

P'Q'R' = true strike direction, N 45 °E – S 45 °W

OS' = true dip amount = 1.42 units

OT = N 70 °E, *OT'* = 3.45 units

Scale 2 cm. = 1 unit

Step II Estimation of Depth in Borehole D

Draw a line OT in Fig. 4.1.1. due N 70 °E, because it is necessary to find out the amount of apparent dip along CD i.e., the bearing between boreholes C and D. The line OT cuts the strike line at T'. The distance OT' is found to be 3.55 units, and therefore the apparent dip is 1 : 3.55. Also the line OT cuts the strike line and therefore it is in the dip direction and hence the formation is to be expected to reach at a level lower than that encountered in borehole C. The distance between the two boreholes is 200 meters and the apparent dip is 1 : 3.55. Therefore the amount of lowering is by 56.3 meters. But in borehole C it is already at a depth of 900 meters, and hence the expected level in borehole D will be 900 + 56.3 i.e., 956.3 meters.

Step III Determining Throw and Kind of Fault

The expected depth of the formation in borehole D is calculated as 956.3 meters. As per the data the formation is actually encountered at a depth of 200 meters only which is a shallow level. Therefore borehole D is located on the *upthrow side* and the amount of throw is 756.3 meters. It is customary to note downthrow side. Therefore the western side is the down throw side of the fault $F_1 - F_1$.

The fault is trending N 10 °E – S 10 °W, while the trend of the formation is N 45 °E – S 45 °W. The fault $F_1 - F_1$ is thus oblique to the strike direction of the formation. Hence it is an oblique fault.

It is possible to classify the fault further. The true dip direction is due S 45 °E, while the down throw is on the western side of the fault i.e., against the direction of dip of the formation. Such fault therefore gets classified as *reverse fault*. This analysis is shown schematically in Fig. 4.1.2.

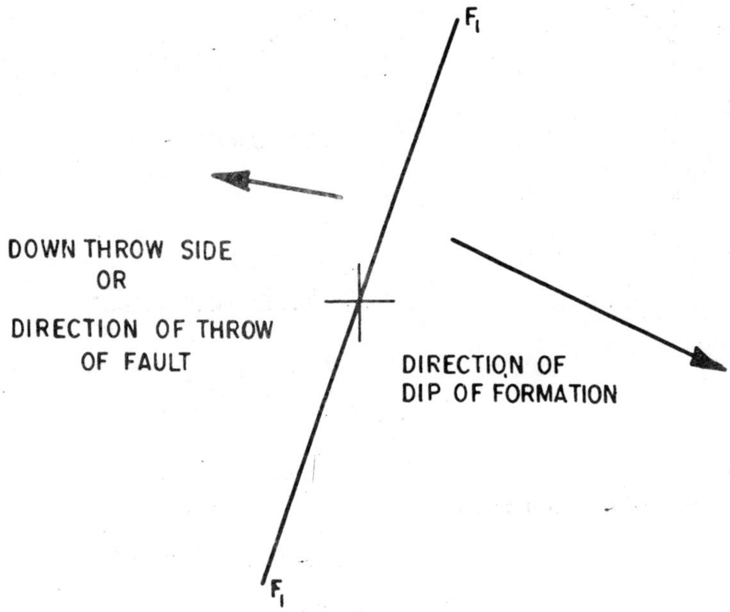

Fig. 4.1.2.

Fig. 4.1.2 Schematic Representation of Direction of Throw of Fault $F_1 - F_1$ and Direction of True Dip of Formation (Problem 4.1)

NOTE : It is a reverse fault because the direction of throw and the direction of true dip are opposite to each other.

In the example described above, borehole D turned out to be on the upthrow side. But in other cases it could as well be on the down throw side. In such an event, the same fault gets classified as a "normal fault", because the direction of dip of the formation and the direction of throw become nearly identical. Therefore in every case the data should be carefully and critically analysed and then the down throw side and the amount of down throw and the kind of the fault is to be established. *In the case of perfect dip faults, the distinction, into normal or reserve faults is not possible.*

4.1.2 Case II Two Faults

This situation increases the varieties of faults like graben, horst and step, considering only the down throw sides. On comparing the down throw direction and that of the dip of the formation, it is possible to recognise reverse, normal or a combinatiion of faults. Illustrative cases are described below.

4.1.3 Parallel Faults (a) *Graben*

Problem 4.2 Boreholes are sunk to ascertain the nature of faults. Borehole A reaches the top of the formation at a depth of 500 meters. Borehole B is 400 meters S 20 °E of A and here it reaches the top at a depth of 400 meters. Borehole C is 600 meters N 70 °E of borehole B and here it reaches the top at a depth of 700 meters. Fault F_1 -F_1 trends E - W and it exists to the immediate north of borehole A, fault F_2 - F_2 also trends E - W but it exists to the immediate south of borehole B. Borehole D is located 500 meters to the north of borehole A and the top is reached at a depth of 100 meters. Borehole E is located 500 meters S 20 °W of borehole B, and here it reaches the top of the formation at a depth of 50 meters. Draw a plan of the boreholes and of the faults to the scale of 1 cm = 100 meters, and determine the structure of the formation and also the nature of the faults. The ground surface is flat.

Procedure

Step I Plan of the Boreholes, the Faults, and Determination of Distance and Bearing between Boreholes A and C (Fig. 4.1.3)

Step II Calculation of Apparent Dips (Only Boreholes A, B, and C are to be considered)

(i) apparent dip along BA

i.e., due N 20 °W $= \dfrac{400}{(500-400)} = 1:4$

(ii) apparent dip along BC

i.e., due N 70 °E $= \dfrac{600}{(700-400)} = 1:2$

(iii) apparent dip along AC

i.e., due S 77°E $= \dfrac{730}{(700-500)} = 1:3.65$

(b) Plotting of apparent dips obtained between boreholes A, B and C

From point O draw lines OP, OQ and OR due N 20 °W, N 70 °E and S 77 °E, respectively, which in turn correspond to directions BA, BC and AC, respectively. Mark off apparent dip amounts of 4, 2 and 3.65 along OP, OQ and OR, and obtain points P', Q' and R'. Join points P', Q', and R' by a straight line. Draw OS perpendicular on $P'Q'R'$ cutting it at S'. OS gives the true dip direction, and OS' gives the amount of true dip. It is found to be 1 : 1.8 and dip direction being N 43 °E (Fig. 4.1.4)

Step III (a) Determination of Throw of Fault F_1 -F_1

The apparent dip along AD i.e., due north is obtained by drawing a long OT due north in Fig. 4.1.4. It cuts the strike line T' and OT' yields the amount of apparent dip which is found to be 1 : 2.55. Since the traverse from borehole A to D is in

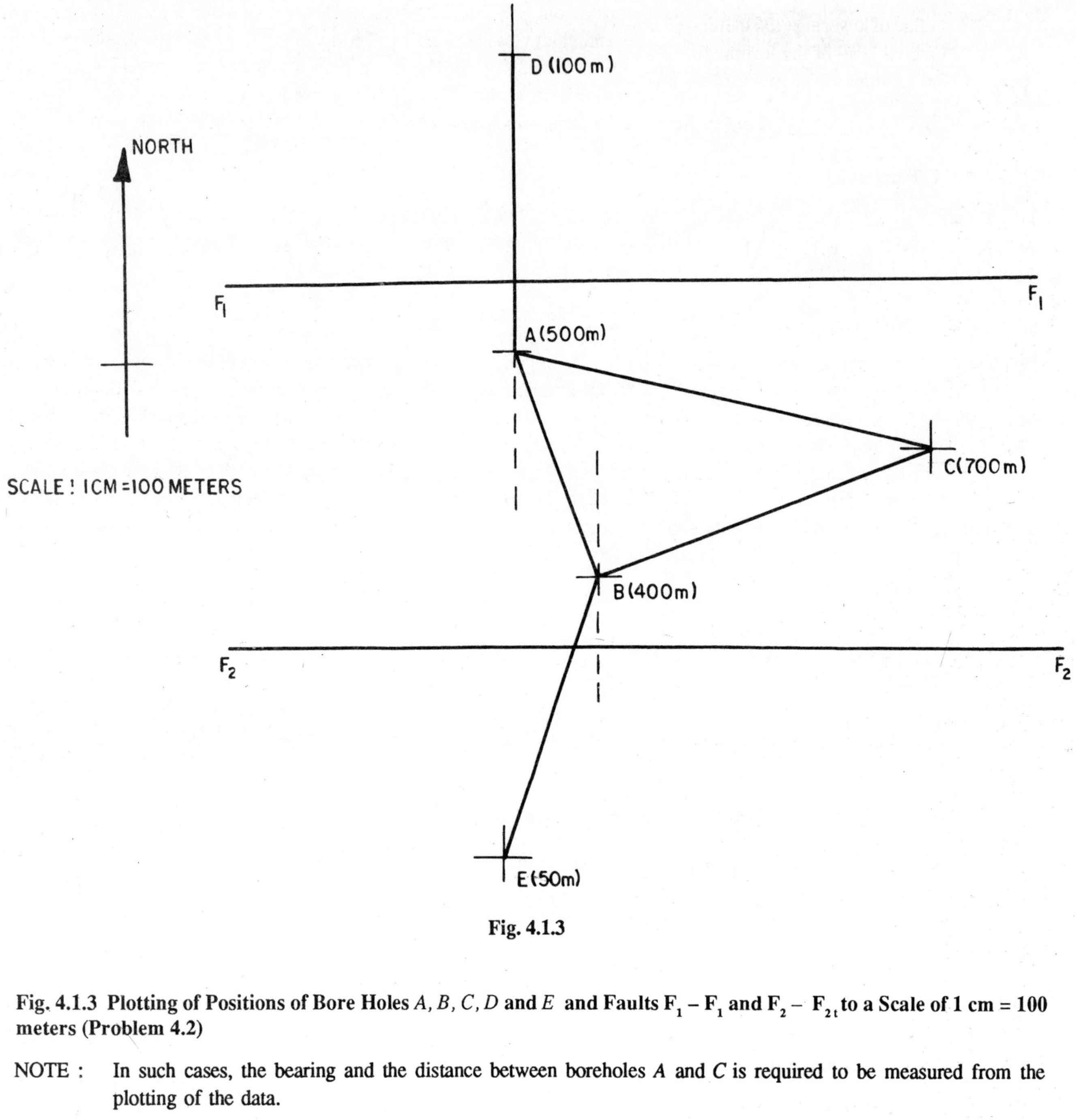

Fig. 4.1.3

Fig. 4.1.3 Plotting of Positions of Bore Holes *A, B, C, D* **and** *E* **and Faults $F_1 - F_1$ and $F_2 - F_2$, to a Scale of 1 cm = 100 meters (Problem 4.2)**

NOTE : In such cases, the bearing and the distance between boreholes *A* and *C* is required to be measured from the plotting of the data.

the (apparent) dip direction, and not in the "anti apparent dip" direction, the formation will be encountered at a level lower than that met with in borehole *A*. The distance between the boreholes *A* and *D* is given as 500 meters. The amount of lowering of the formation over this distance of 500 meters is obtained by the following equation.

$$\text{Lowering between boreholes } A \text{ and } D = \frac{500 \times 1}{2.55} = 196 \text{ meters}$$

This value of lowering (196 meters) is to be added to the level of 500 meters at which the formation is found in borehole *A*. The expected level of the formation in the borehole *D* therefore works out as 500 + 196 = 696 meters. However as per data, the formation is encountered in borehole *D* at a depth of only 100 meters. Hence the formation is thrown up on

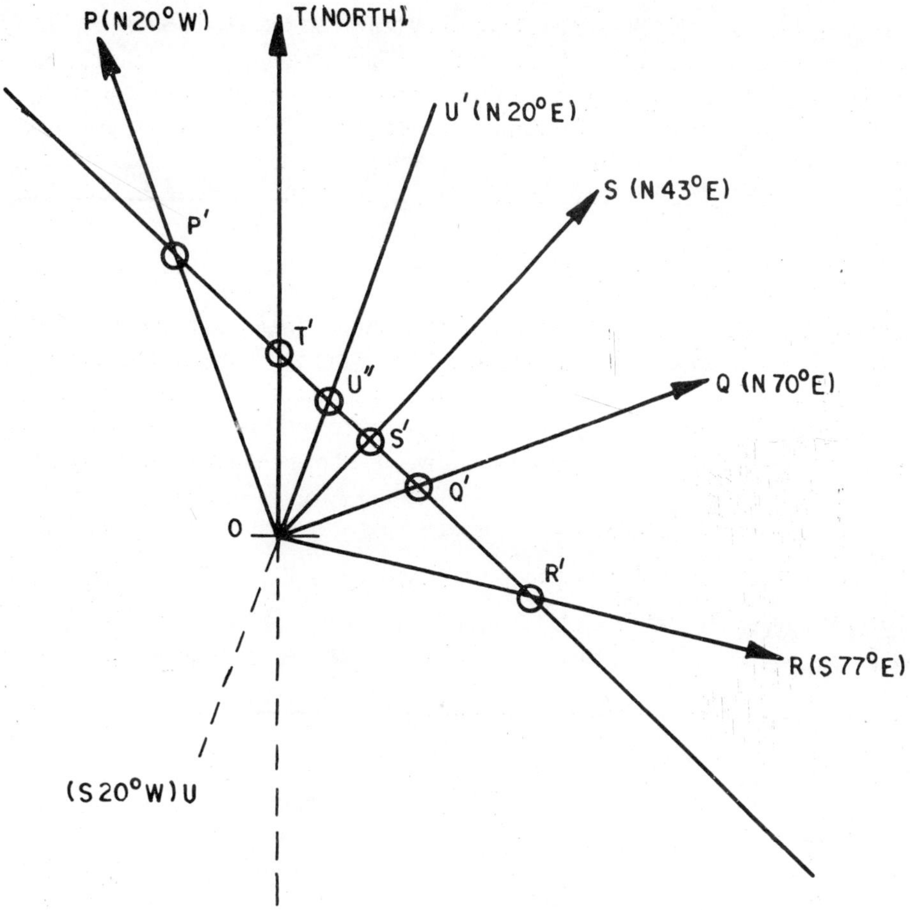

Fig. 4.1.4.

Fig. 4.1.4 Plotting of Apparent Dips Calculated from Bore Holes *A*, *B*, and *C* (Problem 4.2)

OP = N 20 °W, *OQ* = N 70 °E, *OR* = S 77 °E

OP' = 4 units, *OQ'* = 2 units, *OR'* = 3.65 units

OS = true dip direction, N 43 °E

P'Q'R'= true strike direction, N 47 °W - S 47 °E

OS'= true dip amount = 1.8 units

OT = apparent dip along *AD* = north, *OT'* = 2.55 units

OU = apparent dip along *BE* = S 20 °W = anti apparent dip direction

OU' = anti apparent dip direction = N 20 °E

OU" = anti apparent dip amount = 2 units

Scale 1 cm = 1 unit

the northern side of the fault $F_1 - F_1$ by 596 meters (696 minus 100 meters). In other words it means that the *southern side of the fault* $F_1 - F_1$ *is downthrown, and the amount of down throw is 596 meters.*

(b) Determination of Throw of Fault F_2 - F_2

The apparent dip along *BE* i.e., due S 20 °W is to be obtained by drawing a line *OU* as shown in Fig. 4.1.4. However it is found that such a line cannot cut the strike line and hence the same line is extended in just the opposite direction i.e., due N 20 °E (line *OU′* in Fig. 4.1.4). This indicates that the movement from borehole *B* to *E* is not in the (apparent) dip direction, but it in the (apparent) "anti dip direction". The formation therefore gets raised to the surface on moving from *B* to *E*. The amount of apparent dip along *OU′* is found to be 1 : 2. The distance between the boreholes *B* and *E* is given as 500 meters. The amount of "raise" or "shallowing" effected over this distance is calculated by solving the following equation.

$$\text{Amount of raise from } B \text{ to } E = \frac{500 \times 1}{2} = 250 \text{ meters}$$

The depth in borehole *B* is given as 400 meters, and from this value 250 are to be *subtracted* as *BE* marks the "anti apparent dip direction". Thus the expected level in borehole E works out as 150 meters (400 minus 250). But as per the data, the actual depth in borehole E is only 50 meters. Therefore the amount of throw is 100 meters. (150 minus 50). Since the expected level is more than the acutal one, borehole E is located on the upthrow side of the fault $F_2 - F_2$. In other words it means that the northern side of the fault marks the down throw side, the amount of throw being 100 meters.

Step IV Classification of Faults $F_1 - F_1$ and $F_2 - F_2$

The trend of the two faults is E – W, while that of the dipping formation is N 47 °W – S 47 °E. Therefore both the faults get

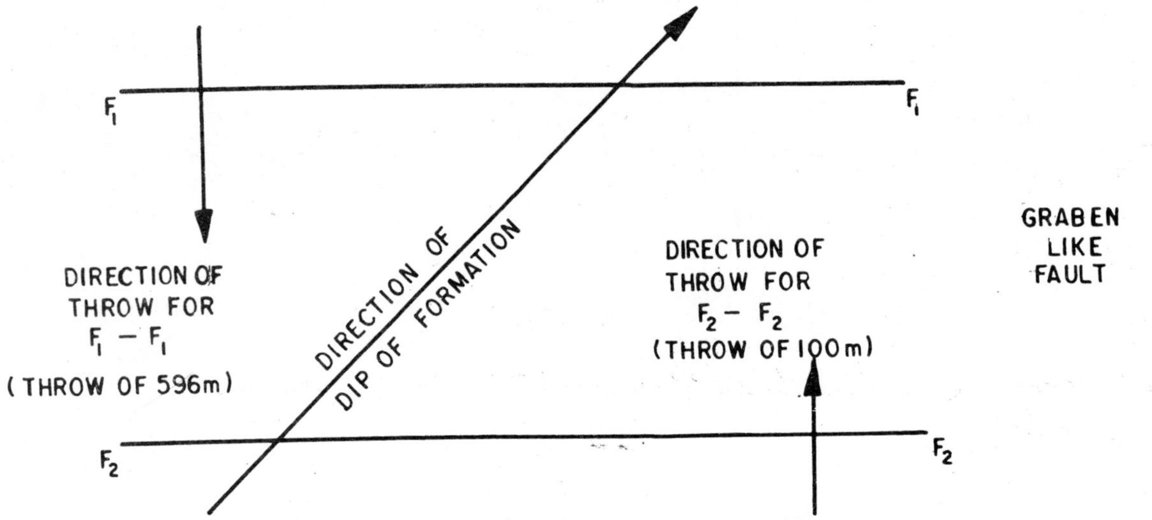

Fig. 4.1.5

Fig. 4.1.5 Schematic Representation of Direction of Throw for the Faults $F_1 - F_1$ and $F_2 - F_2$, and the Direction of Dip of the Formation (Problem 4.2)

NOTE : Fault $F_1 - F_1$ is a reverse fault because the direction of throw of fault (south) is opposite to the direction of true dip of the formation (N 43 °E). Fault $F_2 - F_2$ is a normal fault because the direction of throw of the fault (north) is in the same direction of true dip of the formation (N 43 °E). The two faults together are parallel to each other and the central block has sunk to different amounts. Therefore it is a "graben like" fault.

classified as "oblique faults". The direction of the dip for the formation is due NE (N 43 °E). The direction of throw for the fault $F_1 - F_1$ is due south, which in a generalised manner may be considered as against or opposite of the direction of dip of the formation. The fault $F_1 - F_1$ therefore gets further classified as "reverse fault". In the case of fault $F_2 - F_2$, the direction of throw is due north, and the formation also is dipping generally in that direction (N 43 °E). Therefore the direction of throw and the direction of dip of the formation is indentical for fault $F_2 - F_2$. This fault therefore gets classified as "normal fault".

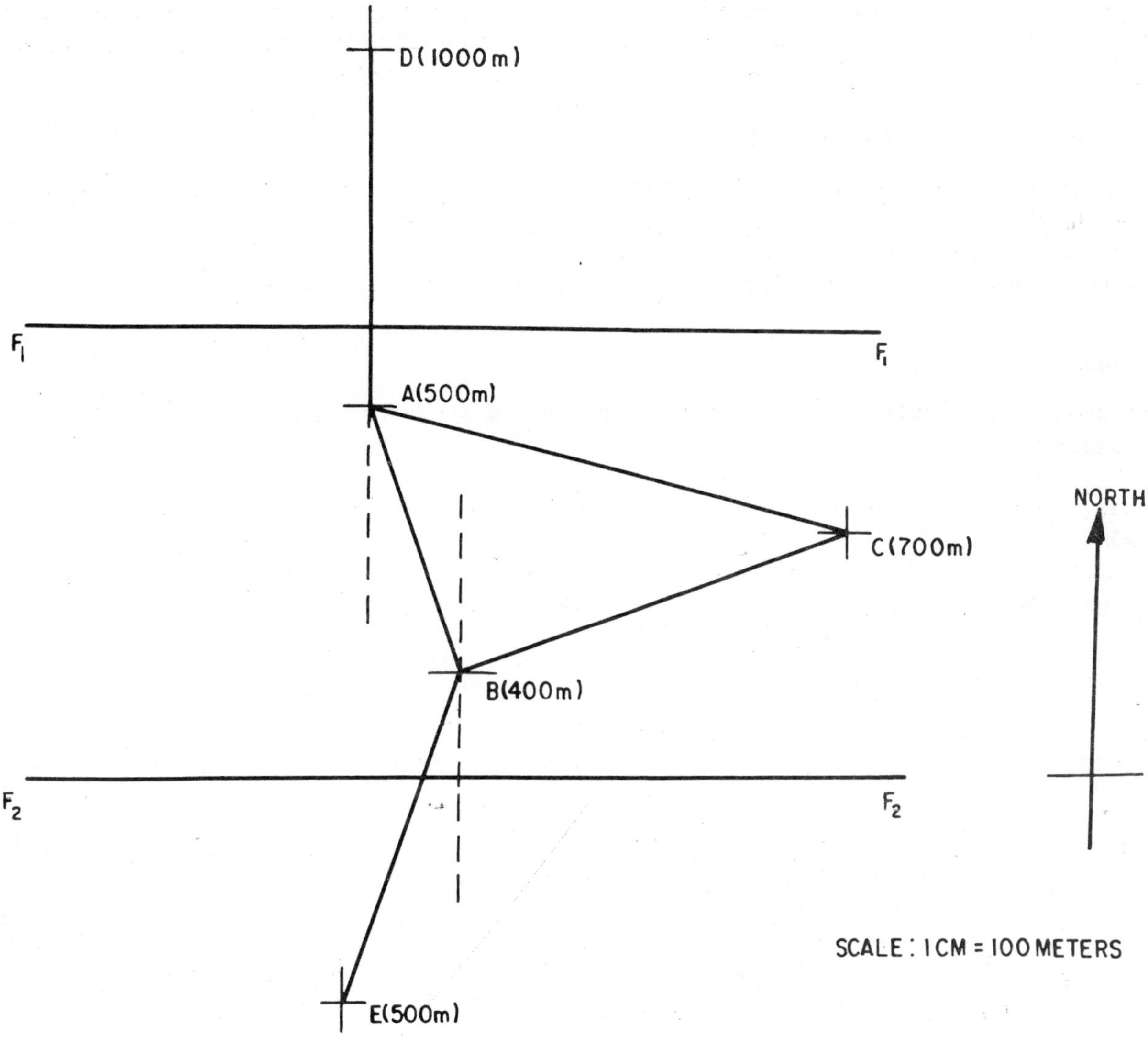

Fig. 4.1.6

Fig. 4.1.6 Plotting of the Positions of Boreholes *A, B, C, D* and *E* and faults $F_1 - F_1$ and $F_2 - F_2$ to a scale of 1 cm = 100 meters ({Problem 4.3)

NOTE : In such cases the bearing and the distance between the boreholes *A* and *C* is to be measured from the plotting of data.

The two parallel faults $F_1 - F_1$ and $F_2 - F_2$ have a common down throw between them (southern side for $F_1 - F_1$ and northern side for $F_2 - F_2$). Therefore the two parallel faults together belong to the category of a "garden. However as the amounts of throw are different (596 meters for $F_1 - F_1$ and 100 meters for $F_2 - F_2$), the fault should therefore be designated as "graben like fault". All these features are schematically shown in Fig. 4.1.5.

4.1.3 (b) Horst

Problem 4.3 Boreholes are sunk to ascertain the nature of faults. Borehole *A* reaches the top of the formation at a depth of 500 meters. Borehole *B* is 400 meters S 20 °E of borehole *A*, and here it reaches the top at a depth of 400 meters. Borehole *C* is 600 meters N 70 °E of borehole *B*, and here it reaches the top of the formation at a depth of 700 meters. Fault F_1 - F_1 trends E – W and is located to the immediate north of borehole *A*. Fault $F_2 - F_2$ also trends E – W, but is located to the immediate south of borehole *B*.

In borehole *D* which is located 500 meters north of borehole *A*, the depth of the formation was found to be 1000 meters. In borehole *E* which is located 500 meters S 20 °W of borehole *B*, the depth of the formation was found to be 500 meters. Draw a plan of the boreholes and the faults to a scale 1 cm = 100 meters and determine the structure of the formation and determine the throw and the nature of the faults. The ground surface is flat.

Procedure

The procedure as outlined in the previous case is to be followed. At the time of synthesis, the inferences drawn will be found to be different.

Step I **Drawing plan of boreholes, the faults and determining the distance and bearing between boreholes *A* and *C* (Fig. 4.1.6).**

Step II **Calculation of apparent dips considering boreholes A, B and C**

(i) apparent dip along *BA*

i.e. due N 20°W $= \dfrac{400}{(500-400)} = 1{:}4$

(ii) apparent dip along *BC*

i.e., due N 70°E $\dfrac{600}{(700-400)} = 1{:}2$

(iii) apparent dip along *AC*

i.e., due S 77°E $= \dfrac{730}{(700-500)} = 1{:}3.65$

(b) Plotting of Apparent Dips Obtained between Boreholes A, B and C

From point *O*, draw lines *OP*, *OQ* and *OR* due N 20 °W, N 70 °E and S 77 °E respectively, which in turn correspond to directions *BA*, *BC* and *AC*, respectively. Mark off apparent dip amounts of 4, 2 and 3.65 along *OP*, *OQ* and *OR* and obtain points *P′ Q′* and *R′* respectively. Join points *P′ Q′* and R′ by a straight line which gives the direction of true strike line. Draw OS perpendicular on *P′Q′R′* cutting it at *S′*. *OS* gives the true dip direction, and *OS′* gives the amount of true dip. It is found to be 1 : 1.8, the dip direction being N 43 °E (Fig. 4.1.7).

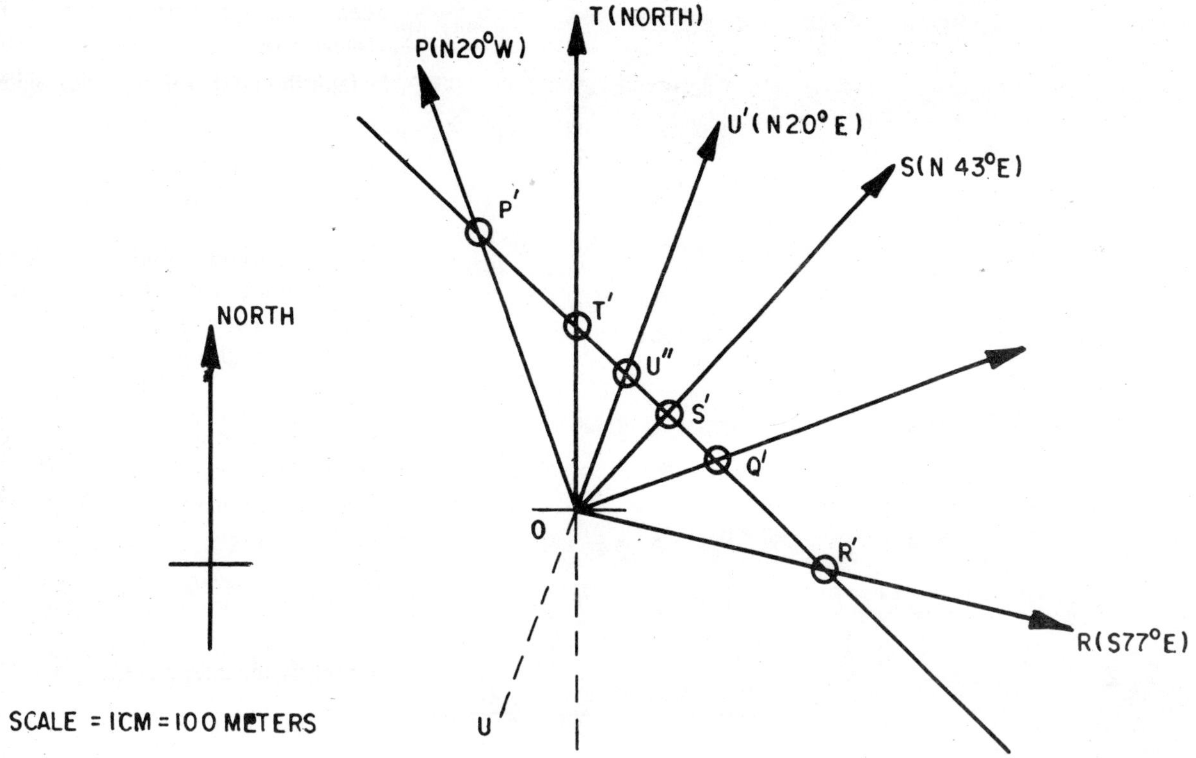

Fig. 4.1.7

Fig. 4.1.7 Plotting of Apparent Dips Calculated from Bore Holes *A, B* and *C* (Problem 4.3)

OP = N 20 °W, *OQ* = N 70 °E, OR = S 77 °E

OP' = 4 units, *OQ'* = 2 units, *OR'* = 3.65 unit

OS = true dip direction, N 43 °E

P'Q'R' = true strike direction, N 47 °W – S 47 °E

OS' = true dip amount = 1.8 units

OT = apparent dip along *AD* = north, *OT'* = 2.5 units

OU = apparent dip along *BE* = S 20 °W,

OU' = anti apparent dip direction = N 20 °E

OU'' = anti apparent dip amount = 2 units

Scale 1 cm = 1 unit

Step III (a) Determination of Throw of Fault F_1 - F_1

The value of apparent dip along *AD* i.e., due north is found to be 1:2.5 (line *OT'* in Fig. 4.1.7). The distance between the boreholes A and D is given as 500 meters, and the direction *AD* is in apparent dip direction, and not in "anti apparent dip" direction. Therefore the amount of lowering over the distance of 500 meters will work out as 200 meters

$\left[\frac{500 - 2 \times 1}{2.5} \right]$. On adding this value of 200 meters to the depth in the borehole *A* (500 meters), a value of 700 meters is obtained, which is the expected level in borehole *D*, if faulting had not taken place. However as per data, tha actual level in borehole *D* is 1000 meters. Therefore borehole *D* is located on the down throw side of the fault, and the amount of throw is 300 meters (1000 minus 700).

(b) Determination of Throw of Fault F2 - F2

The apparent dip along *BE* is found to be 1 : 2 (line *OU″* in Fig. 4.1.7). The distance between the boreholes *B* and *E* is given as 500 meters. However, the direction *BE* is in the anti dip direction (line *OU* in Fig. 4.1.7 cannot cut the strike line, but the line *OU′* which is opposite of *OU*, cuts it). Therefore there will not be lowering of the formation on proceeding from *B* to *E*, but a raise in the level, or shallowing of the formation will take place. The amount of shallowing over a distance of 500 meters works out as 250 meters $\left[\frac{500 - 2 \times 1}{2.5} \right]$. This value is to be subtracted from the level in borehole *B* (400 meters). The expected value of depth in borehole E therefore will be 150 meters (400–250). But the actual level in bore hole B is 500 meters, which is more than the expected one. Therefore borehole *E* is located on the down throw side of the fault and the amount of throw is 350 meters (500 minus 150).

Step IV Classification of Faults F1 - F1 and F_2 - F_2

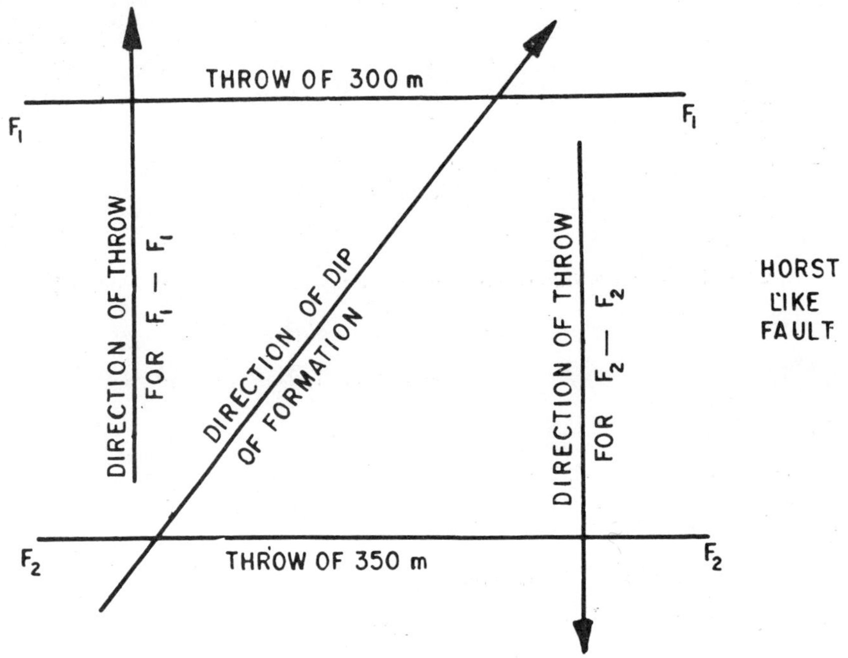

Fig. 4.1.8

Fig. 4.1.8 Schematic Representation of Direction of Throw for the Faults $F_1 - F_1$ and $F_2 - F_2$ and the Direction of Dip of the Formation (Problem 4.3)

NOTE : Fault F1 – F1 is a normal fault because the direction of throw (north) is same as direction of true dip of the formation (N 43 °E). Fault F2– F2 is a reverse fault, because the direction of throw of the fault (south) is against direction of true dip of the formation (N 43 °E). Both the faults are parallel to each other and the central block is thrown up through different amounts. Therefore it is a "horst like " fault.

Both the faults are parallel between themselves, but their trend of E-W is at an angle to the strike direction of the formation

(N 47 °W– S 47 °E). Therefore both the faults are "oblique faults". For fault $F_1 - F_1$, the direction of throw is in the same direction (north) as that of the direction of dip of the formation (N 43 °E). Hence it is a normal fault. For fault $F_2 - F_2$, the throw is (south) against the direction of dip of the formation (N 43 °E). Therefore this fault is a reverse fault. The block in between the two faults is thus observed to be thrown up. Therefore it is a "horst" fault. However the amount of throw being slightly different (300 meters for $F_1 - F_1$ and 350 meters $F_2 - F_2$), it may be classified as "horst-like" fault. These features are presented in the form of a schematic diagram in Fig. 4.1.8.

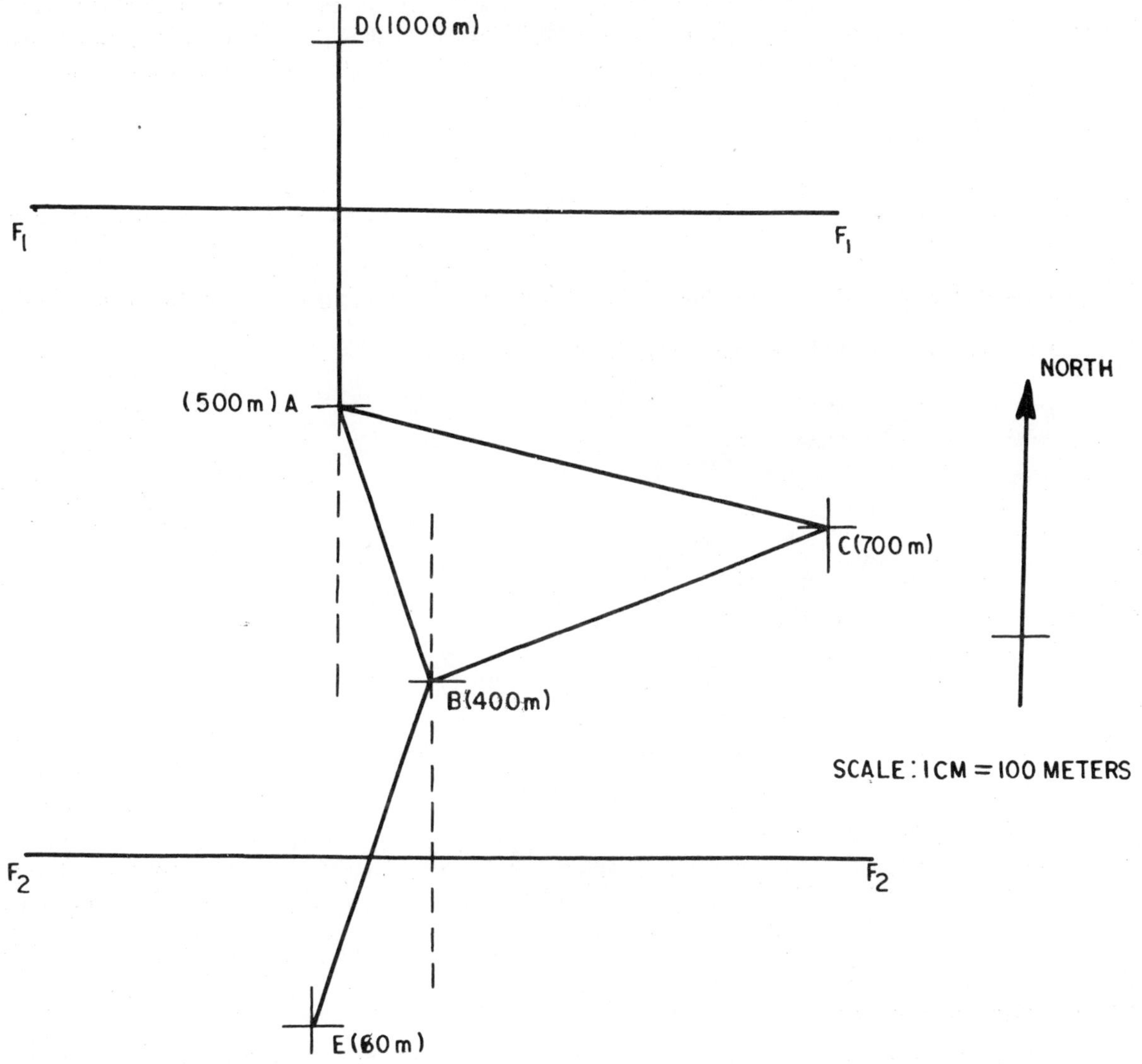

Fig. 4.1.9

Fig. 4.1.9 Plotting of the Positions of Bore holes *A, B, C, D* and *E* and faults $F_1 - F_1$ and $F_2 - F_2$ to a scale of 1 cm = 100 meters (Problem 4.4)

NOTE : In such cases the bearing and the distance between the boreholes A and C are required to be measured from the plotting of data.

4.1.3 Step Fault

Problem 4.4 Boreholes are sunk to ascertain the nature of faults. Borehole A reaches the top of the formation at a depth of 500 meters. Borehole B is 400 meters S20°E of borehole A, and here it reaches the top at a depth of 400 meters. Borehole C is 600 meters N 70°E of borehole B, and here it reaches the top of the formation at a depth of 700 meters. Fault F1-F1 trends E-W and is located to the immediate north of borehole A. Fault F2-F2 also trends E-W but is located to the immediate south of borehole B.

In borehole D which is located 500 meters north of borehole A, the depth of the top of the formation was found to be 1000 meters. In borehole E which is located 500 meters S 20 °W of borehole B, the top of the formation is found at a depth of 50 meters. Determine the attitude of the formation, throw and kind of the faults. Draw a plan of the boreholes to a scale of 1 cm = 100 meters. The ground surface is flat.

Procedure

It is same as that followed in Problem 4.2 and 4.3. The difference lies in the classification of faults. The details are described below.

Step I Drawing plan of boreholes, faults and determining the distance and bearing between boreholes A and C (Fig. 4.1.9).

Step II (a) Calculation of Apparent Dips Considering Boreholes A, B and C

(i) apparent dip along BA

$$\text{i.e., due N } 20^\circ\text{W} = \frac{400}{(500-400)} = 1:4$$

(ii) apparent dip along BC

$$\text{i.e., due N } 70^\circ\text{E} = \frac{600}{(700-400)} = 1:2$$

(iii) apparent dip along AC

$$\text{i.e., due S } 77^\circ\text{E} = \frac{730}{(700-500)} = 1:3.65$$

(b) Plotting of Apparent Dips Obtained from Boreholes A, B and C

From point O, draw lines OP, OQ and OR due N 20 °W, N 70 °E and S 77 °E respectively, which in turn correspond to direction BA, BC and AC, respectively. Mark off apparent dip amounts of 4, 2 and 3.65 along OP, OQ and OR and obtain points P', Q' and R', respectively. Join points P', Q' and R' by a straight line, which gives the direction of true strike line. Draw OS perpendicular $P'Q'R'$ cutting it at S'. OS gives the true dip direction and OS' gives the amount of true dip. It is found to be 1 : 1.8, the dip direction being N 43 °E (Fig. 4.2.0).

Step III (*a*) Determination of Throw of Fault F_1 - F_1

Apparent dip along AD i.e. due north is obtained as 1 : 2.5 (line OT' in Fig. 4.20) and the distance between boreholes A and D is given as 500 meters. The amount of lowering over this distance works out as 200 meters $\left[\frac{500 \times 1}{2.5}\right]$.
The direction AD is in the (apparent) dip direction and not in the "anti apparent" dip direction. Therefore the expected depth in borehole D will be equal to depth in borehole A (500 meters) plus the amount of lowering by 200 meters. This works out as 700 meters. However the actual depth in borehole D is given as 1000 meters which is more than the expected depth of 700 meters. Hence borehole D is located on the down throw side of the fault and the amount of throw is 300 meters (1000 minus 700).

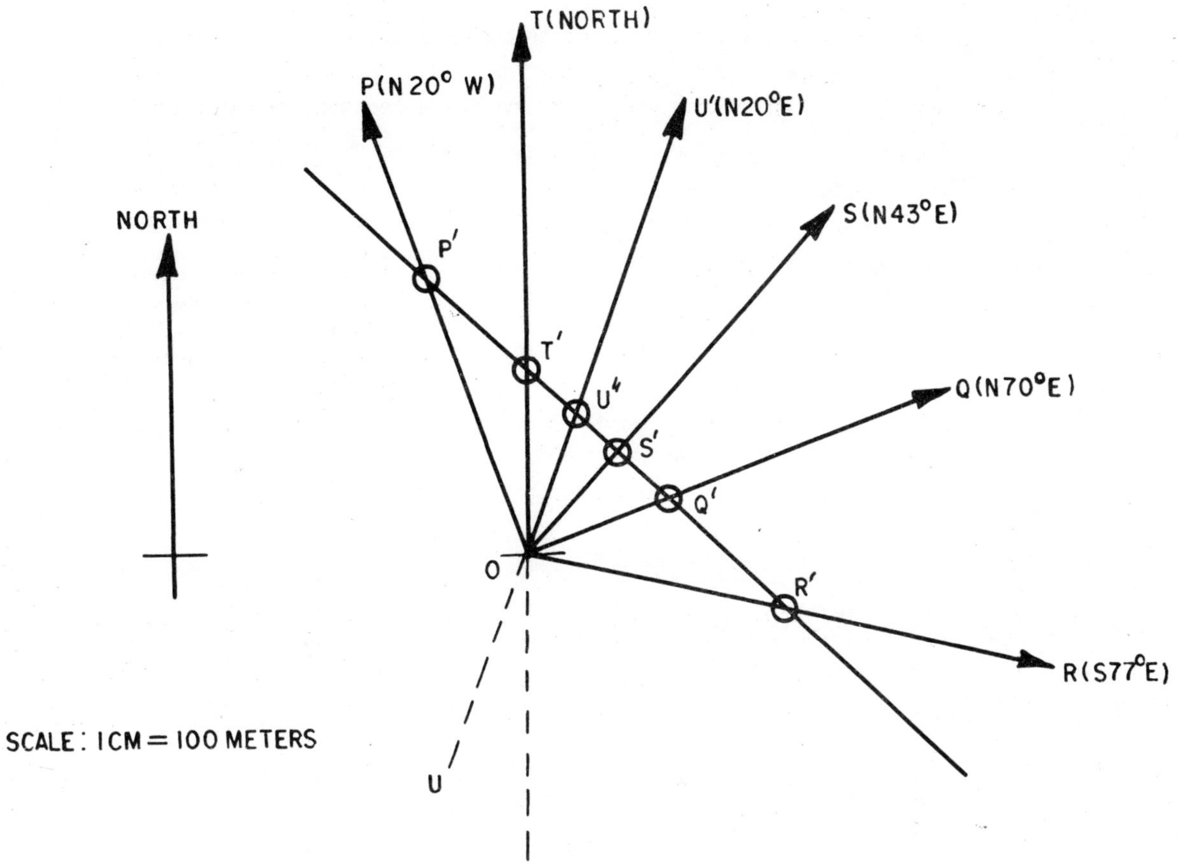

Fig. 4.2.0

Fig. 4.2.0 Plotting of Apparent Dips Calculated from Bore Holes *A.*, *B* **and** *C* **(Problem 4.4)**

OP = N 20 °W, *OQ* = N 70 °E, *OR* = S 77 °E

OP' = 4 units, *OQ'* = 2 units, *OR'* = 3.65 unit

OS = true dip direction, N 43 °E

P'Q'R' = true strike direction, N 47 °W – S 47 °E

OS' = true dip amount = 1.8 units

OT = apparent dip along *AD* = north, *OT'* = 2.5 units

OU = apparent dip along *BE* = S 20 °W,

OU' = anti apparent dip direction = N 20 °E

OU" = anti apparent dip amount = 2 units

Scale 1 cm = 1 unit

(b) Determination of Throw for Fault F2 - F2

Apparent dip along *BE* i.e., due S 20 °W is found to be 1 : 2 (line *OU"* in Fig. 4.2.0), and the distance between the boreholes *B* and *E* is given as 500 meters. Along the direction *BE* (*OU* in Fig. 4.2.0) there will be raise in the level, because this direction is an "anti dip" direction. The amount of raise over a distance of 500 meters is obtained as 250 meters $\left[\frac{500 \times 1}{2}\right]$. The expected level in borehole *E* will be the difference between the depth in borehole *B* and the amount of raise of value 250 meters. It works out as 150 meters (400 minus 250). But the actual depth in borehole *E* is given as only 50 meters, which is much less than the expected level of 150 meters. Hence borehole *E* is located on the upthrow side of tha fault, or the down throw side is located on the northern side of the fault, the amount of throw being of 100 meters.

Step IV Classification of Faults $F_1 - F_1$ and $F_2 - F_2$

Though both the faults are parallel to each other, their E – W trend is at an angle to the strike of the formation. (N 47 °W–S 47 °E). Therefore these faults are "oblique" faults. For fault $F_1 - F_1$ the direction of throw is towards north and the direction of dip of the formation is generally the same (N 43 °E). Hence the fault $F_1 - F_1$ is a normal fault. Fault $F_2 - F_2$ is also a normal fault, because the direction of throw (north) and the direction of dip of the formation (N 43 °E) are same. As the direction of down throw is north for both the faults, this produces a "step –like" structure, and hence the 2 parallel faults are further classified as "step faults". These features are schematically shown in Fig. 4.2.1.

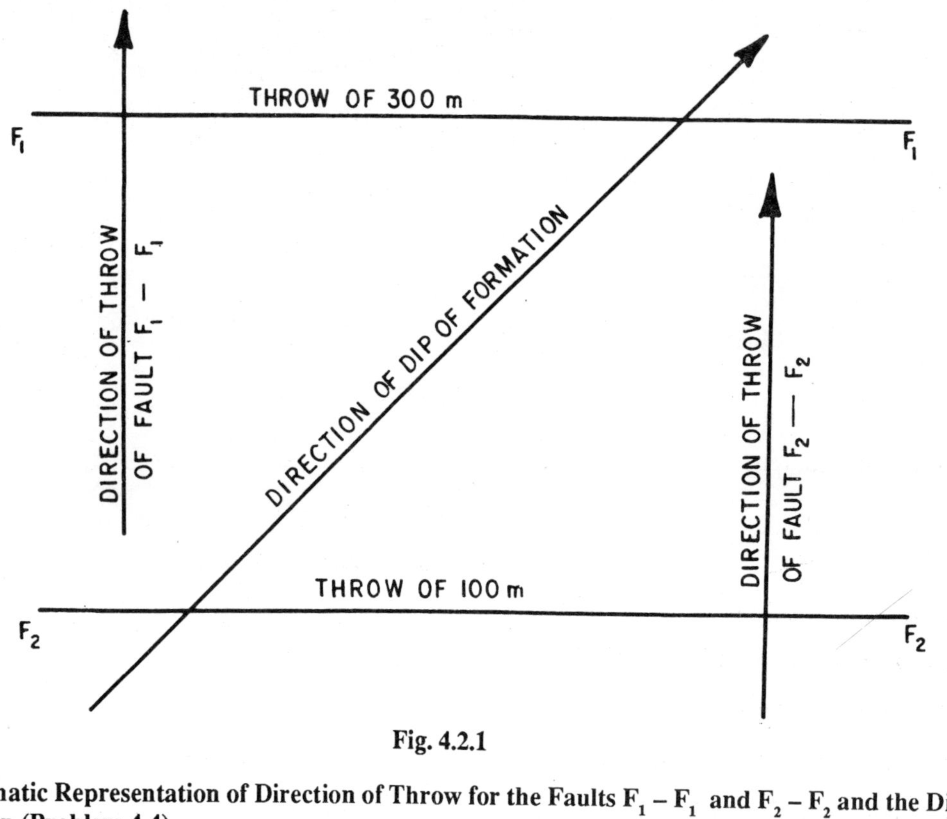

Fig. 4.2.1

Fig. 4.2.1 Schematic Representation of Direction of Throw for the Faults $F_1 - F_1$ and $F_2 - F_2$ and the Direction of Dip of the Formation (Problem 4.4)

NOTE : Fault $F_1 - F_1$ and $F_2 - F_2$ get categorised as normal faults because the direction of throw of the faults (north) is same as direction of true dip of the formation (N 43 °E). Both the faults are parallel to each other and for both of them northern side is the down throw side. Therefore the faults are further classified as "Step" faults.

Fig. 4.1.4 Case III Other Varieties of Faults

Problem 4.5 Boreholes are sunk to ascertain the nature of faults. Borehole *A* reaches the top of the formation at a depth of 500 meters. Borehole *B* is 400 meters S 20 °E of borehole A, and here it reaches the top of the formation at a depth of 400 meters. Borehole C is 600 meters N 70 °E of borehole *B*, and here it reaches the top of the formation at a depth of 700 meters. Fault $F_1 - F_1$ is trending E – W and is located to the immediate north of borehole *A*. Fault $F_2 - F_2$ is trending N 45 °E – S 45 °W and is located to the immediate east of borehole C.

Borehole *D* is located 500 meters north of borehole *A*, and here it reaches the top of the formation at a depth of 100 meters. Borehole *E* is located 500 meters east of borehole *C*, and here it reaches the top of the formation at a depth of 1000 meters. Draw a plan of the boreholes and the faults, to a scale of 1 cm = 100 meters, and determine the attitude of the formation and the kind of the faults. The ground surface is flat.

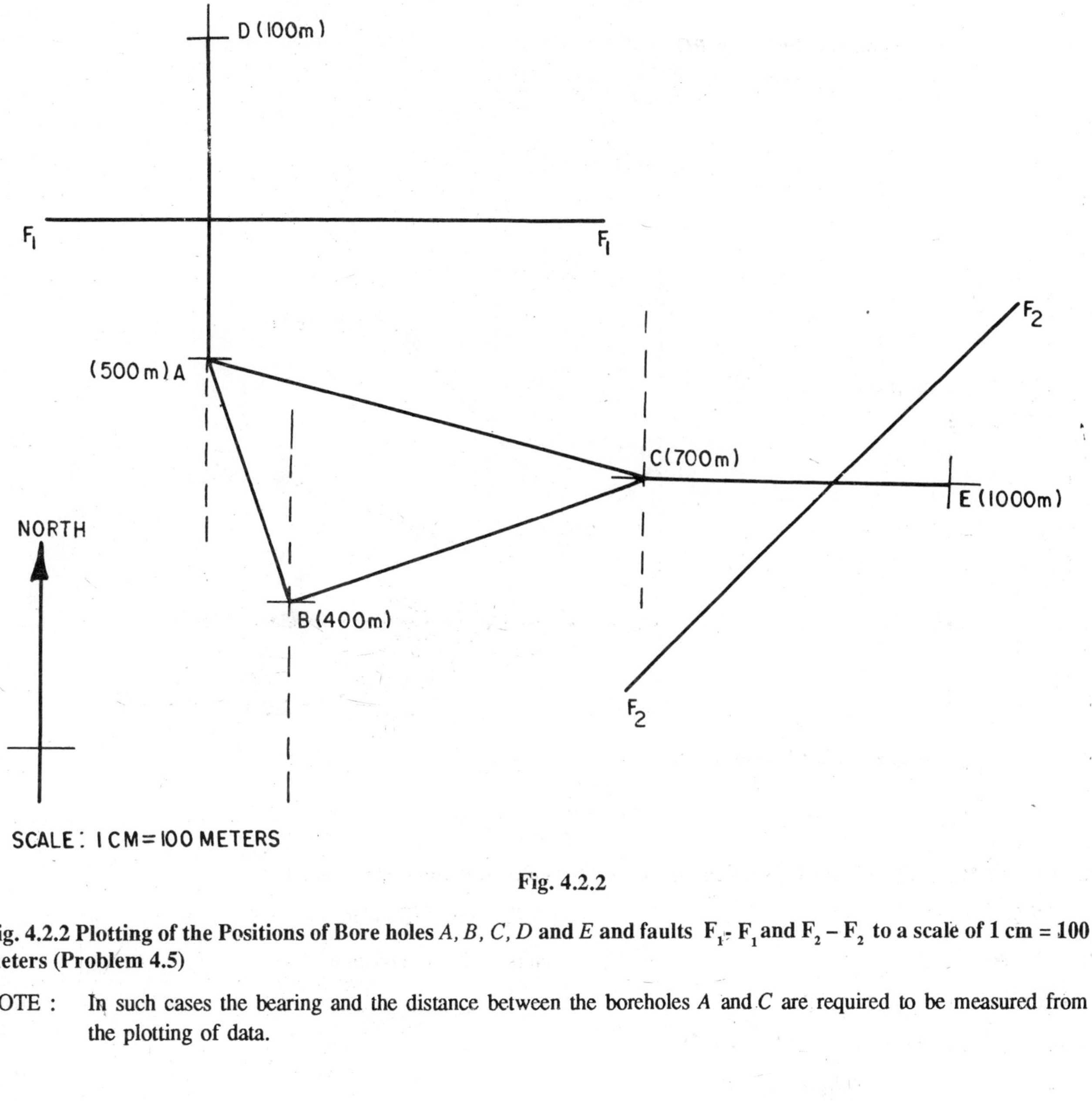

Fig. 4.2.2

Fig. 4.2.2 Plotting of the Positions of Bore holes *A, B, C, D* and *E* and faults F_1- F_1 and $F_2 - F_2$ to a scale of 1 cm = 100 meters (Problem 4.5)

NOTE : In such cases the bearing and the distance between the boreholes *A* and *C* are required to be measured from the plotting of data.

Procedure

It is similar to that followed in Problems 4.2, 4.3 and 4.4. It differs only when the results are synthesized. The details are described below.

Step I Drawing Plan of Borehole, the Fault and Determining the Distance and Bearing between boreholes A and C (Fig. 4.2.2.)

Step II (a) **Calculation of Apparent Dips Considering Boreholes A, B and C**

(i) apparent dip along *BA*

i.e., due N 20°W $= \frac{400}{(500-400)} = 1:4$

(ii) apparent dip along *BC*

i.e., due N 70 °E $= \frac{600}{(700-400)} = 1:2$

(iii) apparent dip along AC

i.e., due 577 °E $= \frac{730}{(700-500)} = 1:3.65$

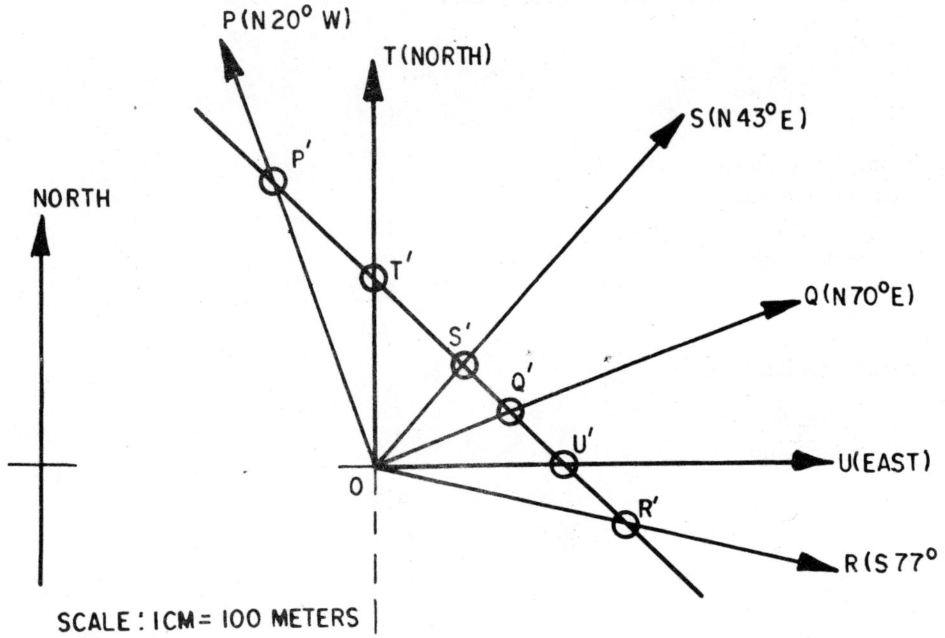

Fig. 4.2.3

Fig. 4.2.3 Plotting of Apparent Dips Calculated from Boreholes *A, B* and *C* (Problem 4.5)

OP = N 20 °W, *OQ* = N 70 °E, *OR* = S 77 °E

OP' = 4 units, *OQ'* = 2 units, *OR'* = 3.65 units

OS = true dip direction, N 43 °E

P'Q'R' = true strike direction, N 47 °W– S 47 °E

OS' = true dip amount = 1.8 units

OT apparent dip along *AD* = north, *OT'* = 2.5 units

OU = apparent dip along *BE* = East, *OU'* = 2.7 units

Scale 1 cm = 1 unit

(iii) apparent dip along AC

i.e., due S 77 °E = $\frac{730}{(700-500)}$ = 1:3.65

(b) Plotting of Apparent Dips Obtained from Boreholes A, B and C

From point O, draw lines OP, OQ and OR due N 20 °W, N 70 °E and S 77 °E, respectively, corresponding to directions BA, BC and AC, respectively. Mark off apparent dip amounts of 4, 2 and 3.65 along OP OQ and OR, and obtain points P', Q' and R' respectively. Join points P', Q' and R' by a straight line, which gives the direction of true strike line. Draw OS perpendicular on $P'Q'R'$ cutting it at S'. OS gives the direction of true dip and OS' gives the amount of true dip. It is found to be 1:1.8, the dip direction being N 43°E (Fig. 4.2.3).

Step III (*a*) Determination of Throw of Fault $F_1 - F_1$

Apparent dip along *AD* is obtained by drawing ling *OT* due north in Fig. 4.2.3. The amount of apparent dip is found to be 1:2.5. The distance between the boreholes A and D is given as 500 meters. The amount of lowering over this distance works out as 200 meters. [$\frac{500 \times 1}{2.5}$]. The expected depth in borehole *D* therefore works out as 700 meters (500 + 200). Value of 200 meters is added to 500, because direction *AD* marks (apparent) dip and not "anti apparent dip". This value of 700 meters is more than the actual value of 100 meters as given in the data for borehole *D*. Therefore the borehole *D* is located on the upthrow side of the fault, and the amount of throw is 600 meters (700 minus 100). In other words it means that the southern side of the fault F_1 - F_1 is the down throw side.

(b) Determination of Throw of Fault $F_2 - F_2$

Apparent dip along *CE* i.e., due east is found by drawing line *OU* in Fig. 4.2.3. It is found to be 1:2.7. This direction *CE* also marks (apparent) dip direction, and not "anti apparent dip direction". Hence the formation will be occurring at a level lower than that found in borehole *C*. The amount of lowering over a distance of 500 meters between borehole *C* and *E* works out as 185.18 meters [$\frac{500 \times 1}{2.7}$]. The expected depth in borehole *E* will work out to 885.18 meters (700 + 185.18). But the actual depth in borehole *E* is given to be 1000 meters, which is considerably more than the expected value. Hence the amount of throw will be 114.82 meters (1000 minus 885.18), and it is on the eastern side of the fault.

Step IV Classification of Faults $F_1 - F_1$ and $F_2 - F_2$

Fault $F_1 - F_1$ trends E-W and the formation has the strike direction of N 47 °W – S 47 °E. Therefore it is an "oblique" fault. Further because the direction of throw of this fault (south) is against the direction of dip of the formation (N 43 °E), it gets categorised as a reverse fault. Fault $F_2 - F_2$ trends N 45 °E – S 45 °W and it nearly coincides with the direction of dip of the formation (N 43 °E). Therefore it is a "dip fault". It cannot be further classified into reverse or normal because the direction of throw is nearly perpendicular to that of direction of dip of the formation, and the distinction cannot be made. The pattern of throw for the faults is such that it is in the same direction, namely south. But because the faults are not parallel to each other, these are not exactly "step faults" but may be classified as "step like" faults. These features are schematically shown **in Fig. 4.2.4.**

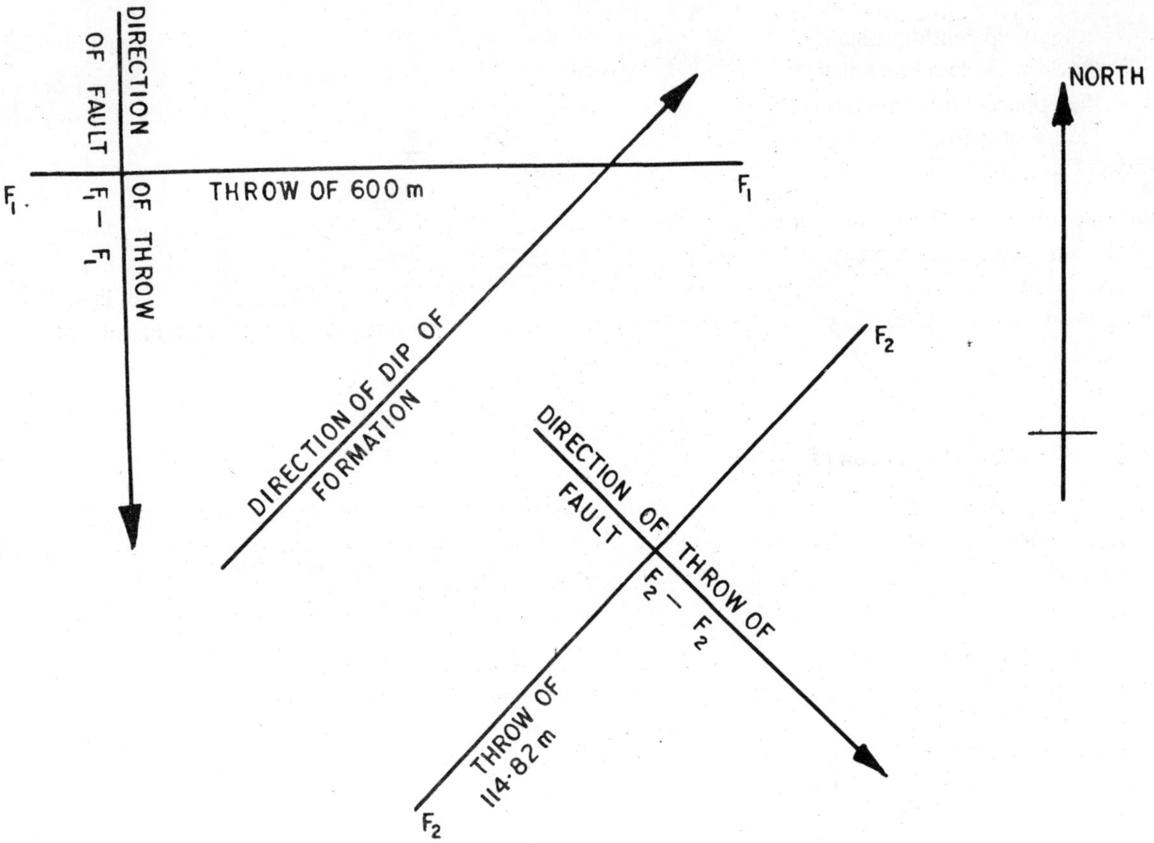

Fig. 4.2.4

Fig. 4.2.4 Schematic Representation of Direction of Throw for the Faults $F_1 - F_1$ and $F_2 - F_2$ and the Direction of Dip of the Formation (Problem 4.5)

NOTE : Fault $F_1 - F_1$ is a reverse fault because the direction of throw (south) is against direction of true dip of the formation (N 43 °E). Fault $F_2 - F_2$ cannot be distinguished into normal or reverse because it is a dip fault. Though faults $F_1 - F_1$ and $F_2 - F_2$ are not parallel to each other, still these get categorised as "step like" faults, because for both of them southern side is the down throw side.

4.1.5 GUIDELINES FOR ASCERTAINING THE DOWN THROW SIDE AND THE AMOUNT OF THROW OF FAULTS

Steps given below are applicable to faults of translation alone.

1. Determine the attitude (true dip direction and amount) of the geologic body by considering the given 3 (or more) boreholes that are occurring on the same side of the fault.

2. Consider one of the boreholes from amongst the 3 (or more) like *A*, *B*, or *C*, and a borehole situated on the other side of the fault, like *D* or *E* (Problem No. 4.1, 4.2, 4.3. 4.4, 4.5).

3. Measure the distance between the two boreholes (like *A* and *D* or *B* and *D* or *C* and *D* in Problem *5.1*; *A* and *E*, or B and *E*, or *C* and *E* in Problem 4.2, 4.3, 4.4, 4.5) and divide the distance by the amount of apparent dip

or anti apparent dip obtained along *AD*, or *BD*, or *CD*, or *AE*, or *BE*, or *CE*. This gives the amount of *lowering* between the two boreholes (if the traverse from *A* to *D*, *B* to *D*, *C* to *D*, *A* to *E*, *B* to *E*, or *C* to *E* boreholes is in the apparent dip directiion). It however gives the amount of *shallowing or raise* between the two boreholes (if the traverse from *A* to *D*, *B* to *D*, *C* to *D*, *A* to *E*, *B* to *E* or *C* to *E* is in the anti apparent dip direction).

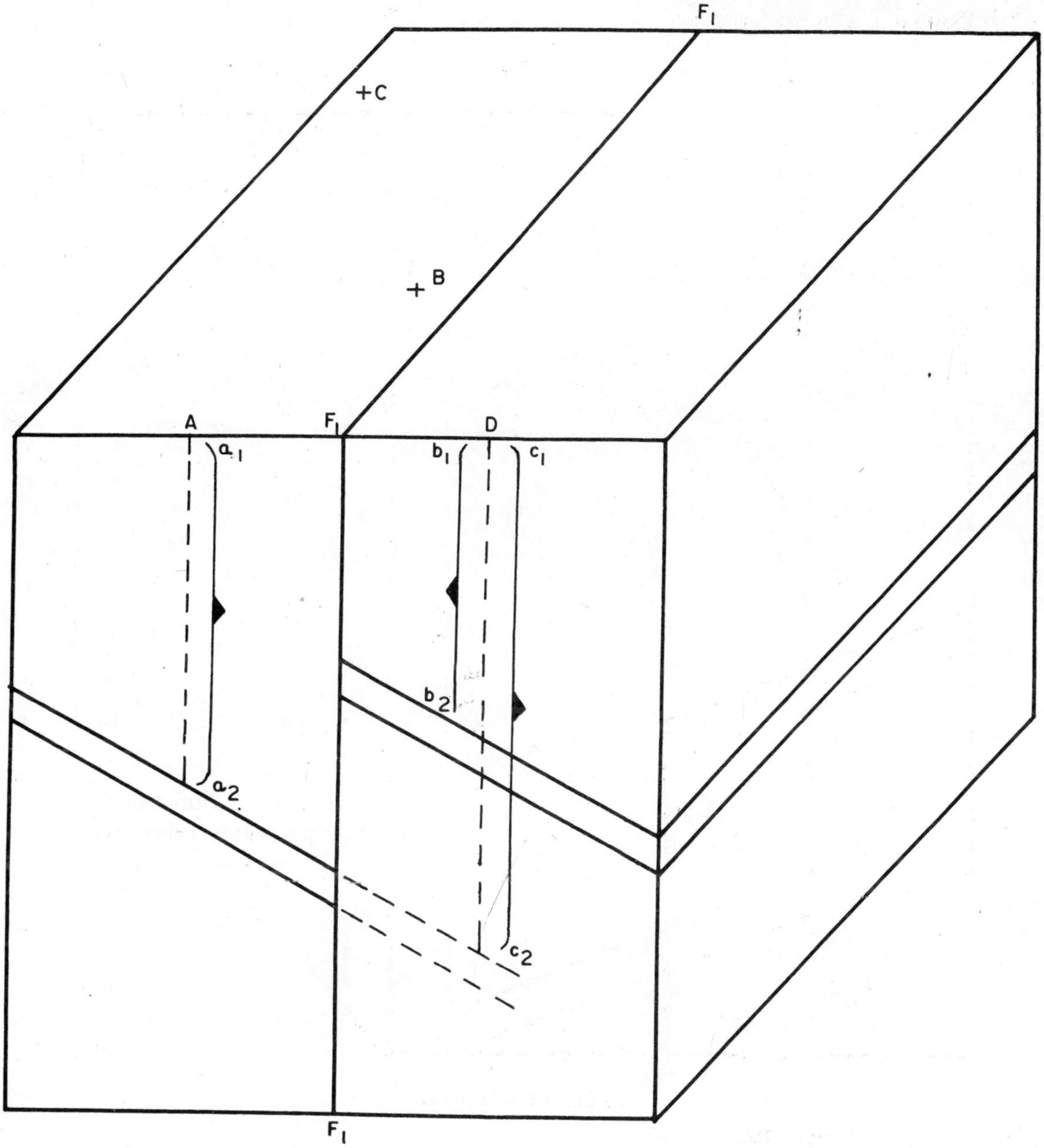

ILLUSTRATION I

ILLUSTRATION I (REVERSE FAULT)

A, *B*, *C*, *D* = sites of boreholes, $F_1 - F_1$ = Fault

$a_1 - a_2$ = given depth of geologic body in borehole *A*

$b_1 - b_2$ = given actual depth of geologic body in borehole *D*

$c_1 - c_2$ = expected (calculated) depth of geologic body in borehole *D*, prior to faulting.

$c_1 - c_2$ (calculated/expected depth) is greater than $b_1 - b_2$ (actual/given depth).

Therefore D is inferred to be located on the upthrow side of fault $F_1 - F_1$. Amount of throw = $c_1 - c_2$ minus $b1 - b2$

NOTE = Value of $c_1 - c_2$ depends upon the value of $a_1 - a_2$, distance between boreholes *A* and *D*, and dip direction and amount of dip of the geologic body. Direction of throw is against the direction of dip of the geologic body. It is therefore is a reverse fault.

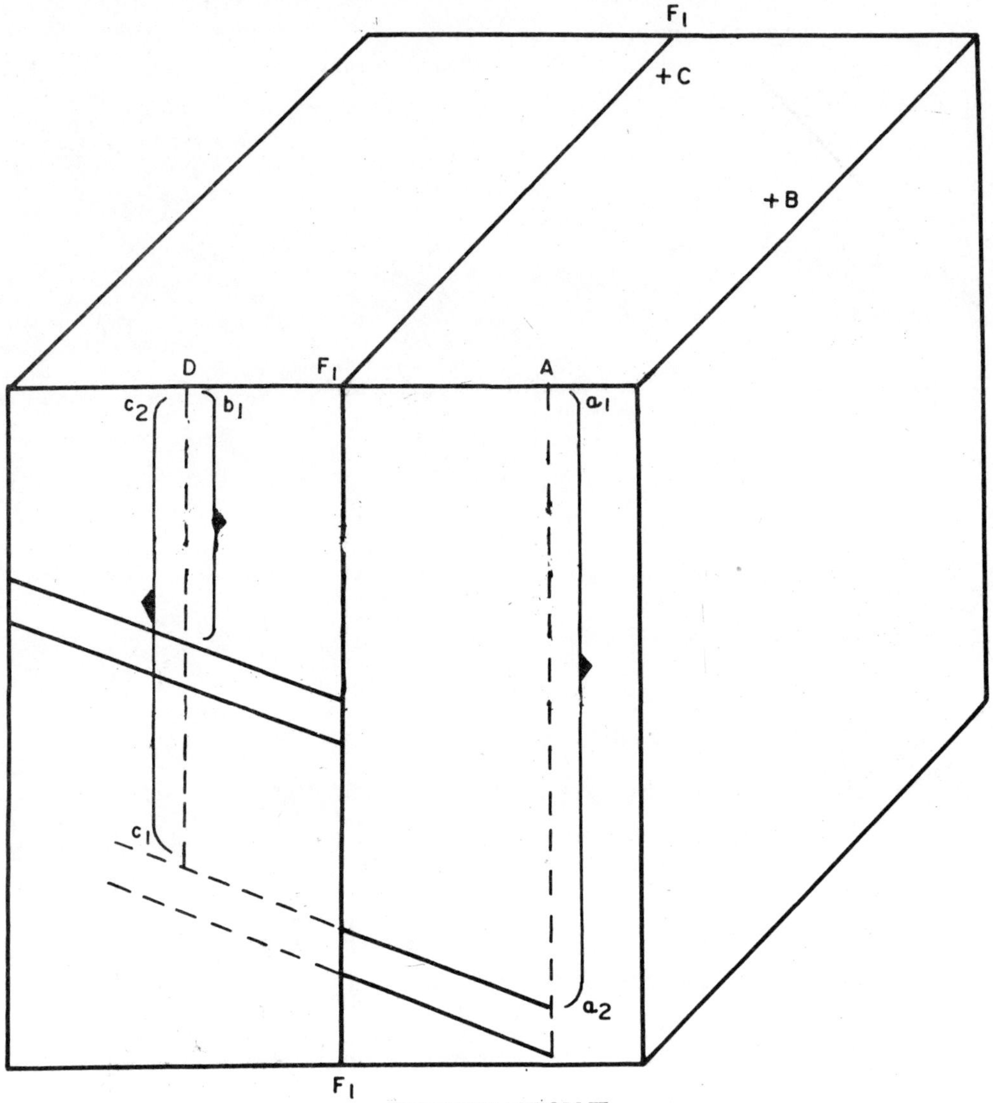

ILLUSTRATION II

ILLUSTRATION II (NORMAL FAULT)

$A, B, C, D =$ sites of boreholes, $F_1 - F_1 =$ fault

$a_1 - a_2 =$ given depth in borehole A

$b_1 - b_2 =$ actual given depth in *D*

$c_1 - c_2 =$ expected (calculated) depth in borehole *D*, prior to faulting.

$c_1 - c_2$ (calculated) > $b_1 b_2$ (actual given depth).

Therefore D is located in the upthrow side of fault F_1-F_1. Amount of throw = c_1-c_2 minus b_1-b_2

NOTE = Value of c_1-c_2 depends upon the value of a_1-a_2, distance between boreholes *A* and *D*, and dip direction and amount of dip of the geologic body. Direction of throw is same as that dip of the geologic body. It is therefore a normal fault.

4. The amount of lowering obtained in step 3 above, is to be added to the level given in the data for the boreholes *A* or B or *C* as the case might be. This procedure gives the value of level expected in boreholes *D* or *E* which are located across the fault. The amount of shallowing or raise obtained in Step 3 above, is to be subtracted from the level given in data for the boreholes *A* or *B* or *C* as the case might be. This procedure gives the value of level expected in boreholes *D* or *E* that are located across the fault.

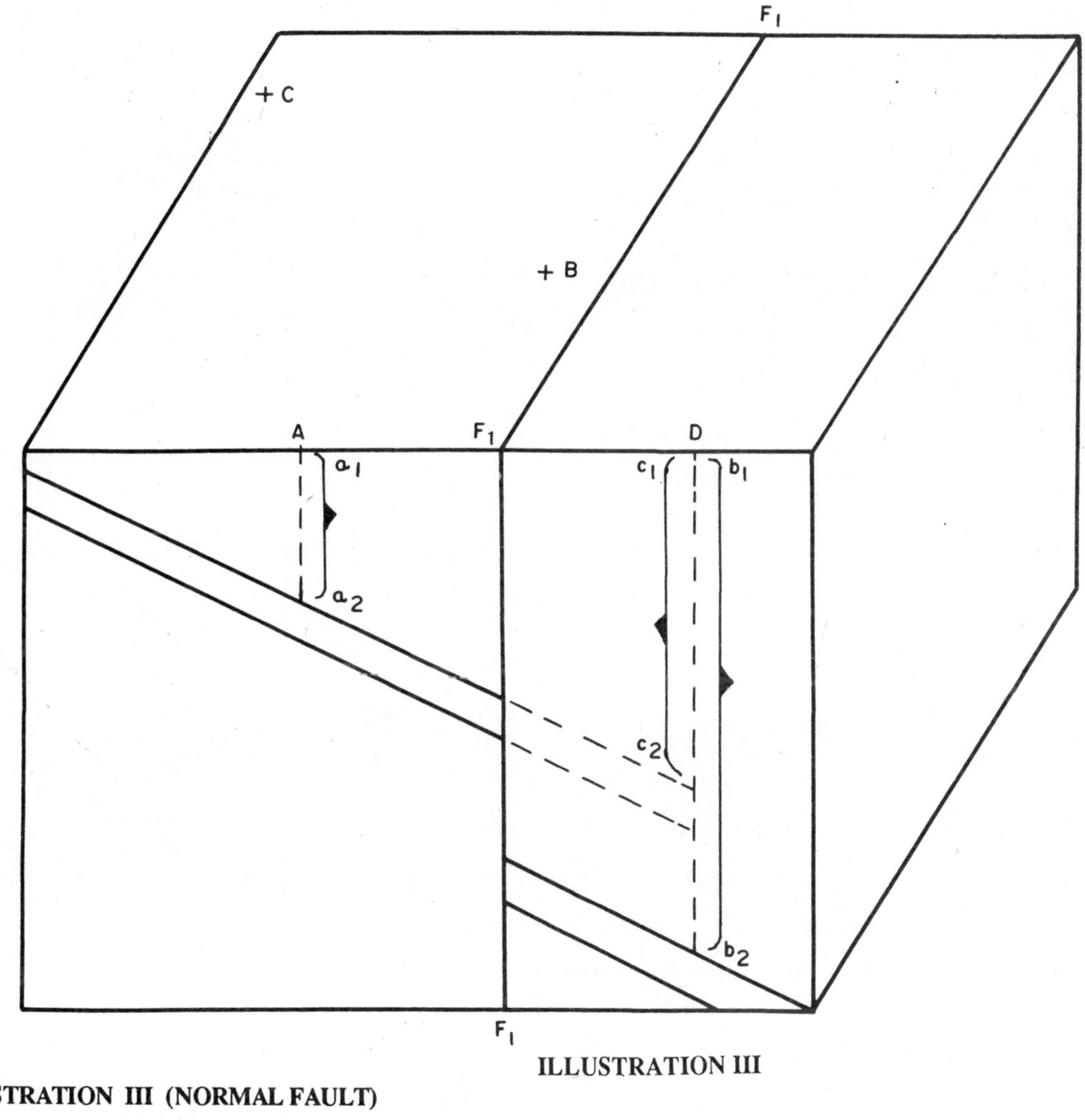

ILLUSTRATION III

ILLUSTRATION III (NORMAL FAULT)

A, *B*, *C*, *D* = sites of boreholes, F_1-F_1 = Fault

a_1-a_2 = given depth in borehole *A*

$b_1 - b_2 =$ actual given depth in boreholee *D*

$c_1 - c_2 =$ expected (calculated) depth in borehole *D*, prior to faulting.

$c_1 - c_2$ (calculated depth) $<$ $b_1 - b_2$ (actual given/depth).

Therefore borehole *D* is located in the upthrow side of fault. Amount of throw = $b_1 - b_2$ minus $c_1 - c_2$

NOTE = Value of $c_1 - c_2$ depends upon the value of $a_1 - a_2$, distance between boreholes *A* and *D*, dip direction and amount of dip of the geologic body. Direction of throw is same as that dip of the geologic body. It is therefore a normal fault.

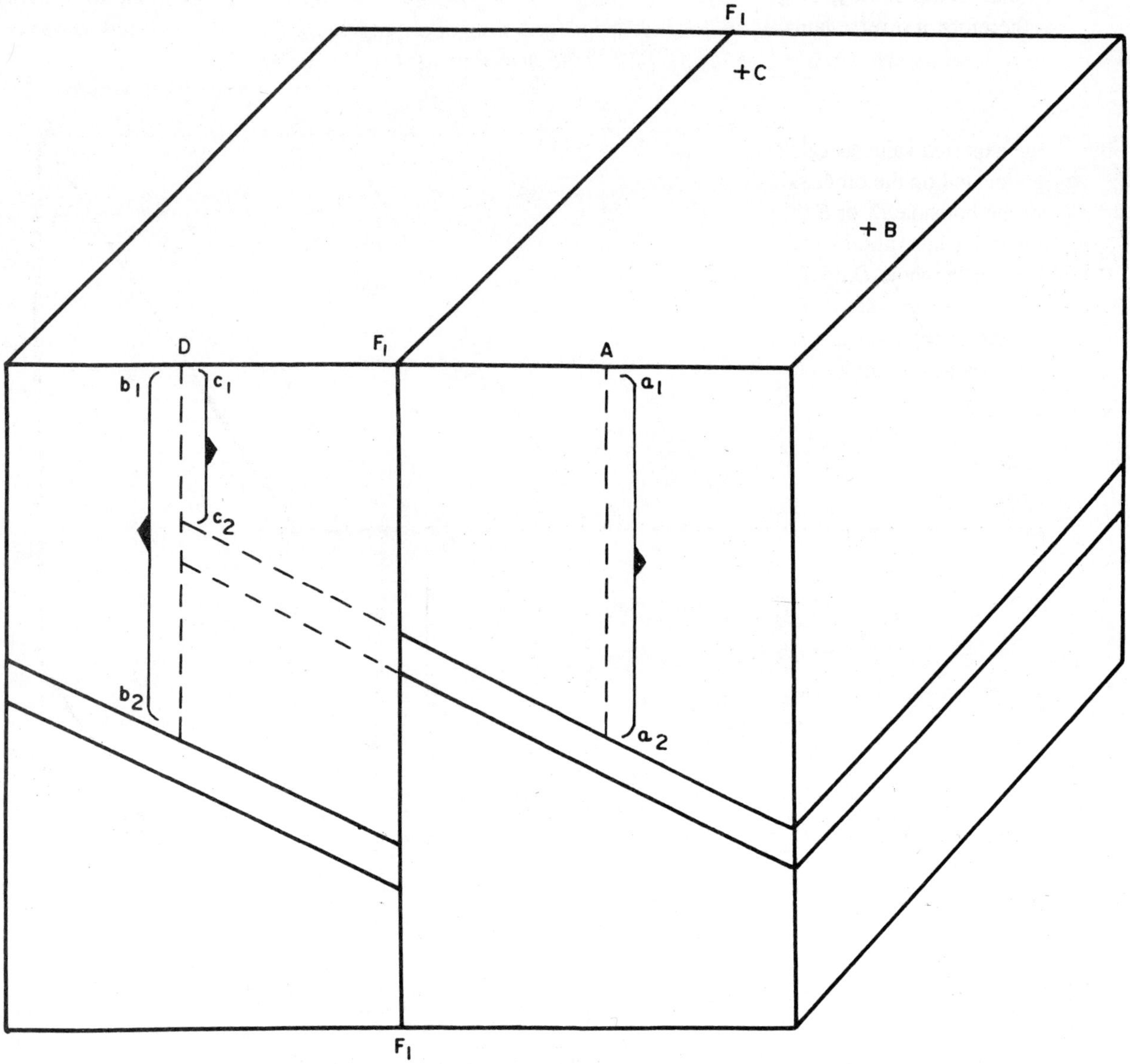

ILLUSTRATION IV

ILLUSTRATION IV (REVERSE FAULT)

A, *B*, *C*, *D* = sites of boreholes, $F_1 - F_1 =$ Fault

$a_1 - a_2 =$ given depth in borehole A

$b_1 - b_2$ = actual given depth in bore hole *D*

$c_1 - c_2$ = expected (calculated) depth in borehole *D*, prior to faulting.

$c_1 - c_2$ (calculated depth) $< b_1 - b_2$ (actual given depth).

Therefore borehole *D* is located on the down throw side of fault $F_1 - F_1$. Amount of throw = $b_1 - b_2$ minus $c_1 - c_2$

NOTE = Value of $c_1 - c_2$ depends upon the value of $a_1 - a_2$, distance between boreholes *A* and *D*, dip direction and amount of dip of the geologic body. Direction of throw is against the direction of dip of the geologic body. It is therefore a reverse fault.

5. The expected value as calculated in step 4 above, is compared with that given in data for boreholes *D* or *E* that are located on the other side of the fault. If the expected value is found to be more than the actual value given for the borehole *D* or *E* then the concerned bore hole (*D* or *E*) is inferred to be located on the upthrow side of the fault (illustration I and II). The difference between the expected level and the actual level as given in the data for borehole *D* or *E*, gives the amount of throw of the fault. In such cases the down throw side is on the side where boreholes *A*, *B* and *C* are situated (Illustration I and II). The difference between the expected level and the actual level as given in the data for borehole *D* or *E*, gives the amount of throw of the fault. In such cases the down throw side is on the side where boreholes *A*, *B* and *C* are situated (Illustration I and II).

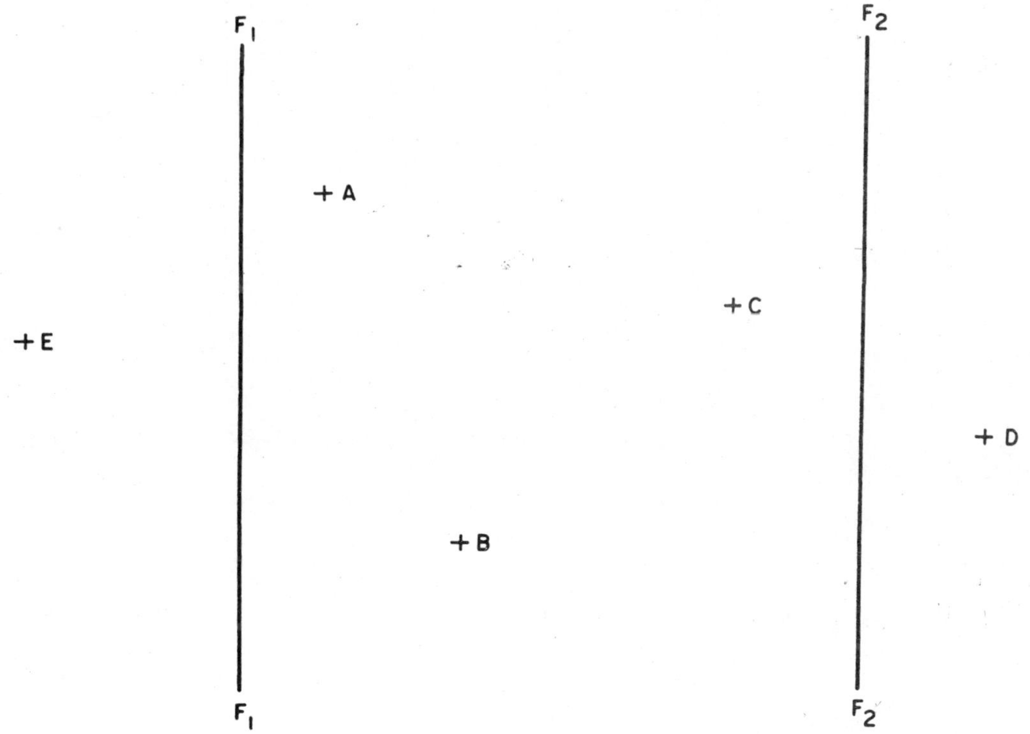

ILLUSTRATION V

ILLUSTRATION V Parallel fault and boreholes *D* and E are located on either sides of boreholes *A*, *B*, *C* as well as the faults $F_1 - F_1$ and $F_2 - F_2$.

If on the other hand the expected value is less than that given for the borehole *D* or *E*, then the borehole *D* or *E* is inferred to be located on the downthrow side of the fault. (Illustration III and IV). The difference between the expected level and actual level as given in the data for borehole *D* or *E*, gives the amount of throw of the fault.

6. The actual position of the boreholes *D* and *E* with respect to the boreholes *A*, *B* and *C*, are of great importance in the case of two or more faults. Thus if the faults be parallel to each other, then in such cases the boreholes *D* and *E* may be located on either sides of boreholes *A*, *B*, and *C*, as shown in Illustration V. Further depending upon the down throw side of the faults, such faults may turn out to be graben (Problem 4.2), horst (Problem 4.3) or step faults (Problem 4.4.)

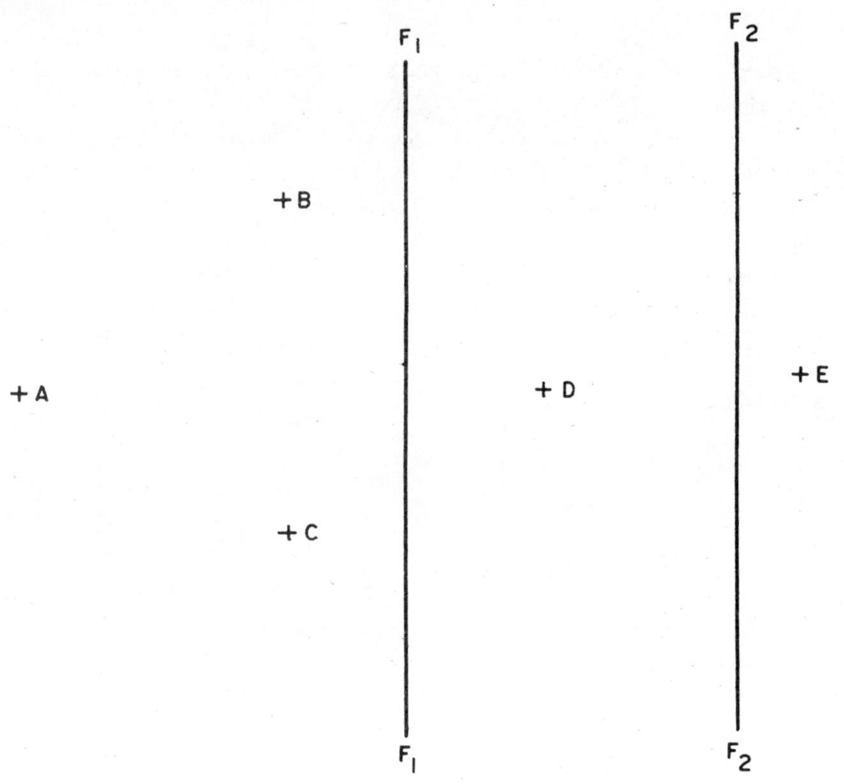

ILLUSTRATION VI

ILLUSTRATION VI Parallel faults but boreholes *D* and *E* are located on right hand side of boreholes *A*, *B* and *C* as well as fault F_1 - F_1 and F_2 - F_2.

The faults may be parallel to each other, but the boreholes *D* and *E* may be situated on only one side of the boreholes *A*, *B*, and *C* as shown in Illustratioin VI. In such cases, the value in boreholes *E* (Illustration VI) gets affected by the effect of fault $F_1 - F_1$ on borehole *D*. Therefore it is necessary to consider this aspect while proceeding to calculate the expected value in borehole *E*.

The two or more faults need not be parallel to each other. Such a situation merely produces a different pattern of faulting. Thus in Illustration VII, borehole *D* is located north of borehole *A*, *B* and *C*, while borehole *E* is located east of boreholes *A*, *B* and *C*.

What is desired to be stressed here is the fact that borehole *D* or *E* may be located in the general dip direction or against the general dip direction of the geologic body. In the former case there is a lowering in the expected level in borehole *D* or *E* with respect to either *A*, *B* or *C* boreholes; whereas in the latter case, there is rase or shallowing in the expected level in borehole *D* or *E*, with respect to borehole *A*, *B* and *C*. It is presumed here that the borehole *D* and *E* are located across the faults with respect to boreholes *A*, *B* and *C*.

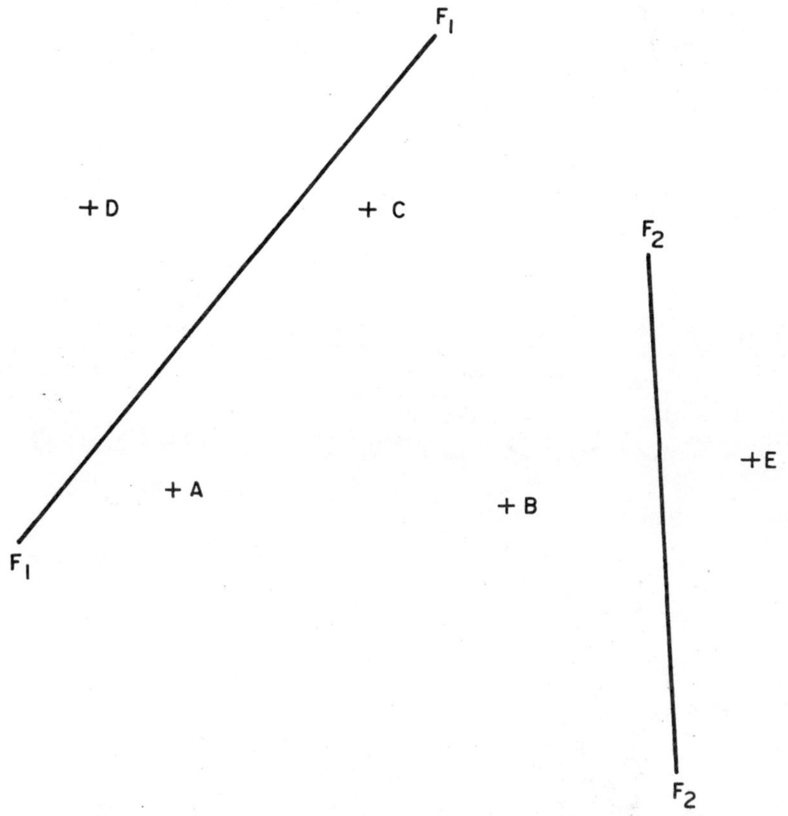

ILLUSTRATION VII

ILLUSTRATION VII Faults are not parallel, yet boreholes are situated on the either sides of the boreholes *A*, *B* and *C*.

5

USE OF STEROGRAPHIC NET IN GEOLOGICAL PROBLEMS

5.1.0 INTRODUCTION

The graphical and the scalar methods described in Chapter I are no doubt useful in determining the true dip, the apparent dip, direction of strike etc. However the procedure involves more time, as lines are to be drawn to scale, angles are to be plotted, proper scale to be chosen for plotting apparent dips and so on. Whenever angular measurements are involved, the sterographic net can be used, and the time taken is considerably less than that taken for the vector and the scalar methods. The use of sterographic net in solving the dip and strike problems, and the determination of the line of intersection of dipping planes will be described in the following pages.

5.1.1. The Stereographic Net

This is derived from a spherical projection which is effected on the equatorial plane. All the angular relations are faithfully retained and hence it is used in the handling of data based on angular measurements. Thus the inclination of surfaces, trend directions, apparent and true dip directions, intersection of dipping surface and the like, can be conveniently reduced to poles or arcs of great circles.

The net consists of equatorial or primitive circle, within which arcs of great circles (which look like longitudes), and meridians (which look like latitudes) are drawn (Fig. 5.1.0). Each arc of great circle or the meridian stands for a value of two degrees. The angular distance therefore between any two consecutive arcs within the net corresponds to two degrees. Therefore the stereographic net which is called as "Wulfe's net", can yield accuracy of only two degrees.

5.1.2 Role of Meridians

These are used to plot directions like north, east S 60 °W, N 30 °E, N 50 °W and so on. The north pole stands for 0°, 360° or North direction. Ends of east, south and west radii mark the east, south and the west directions respectively. The other meridians (latitudes) meet the primitive of the equitorial great circle, and such places mark the several directions like

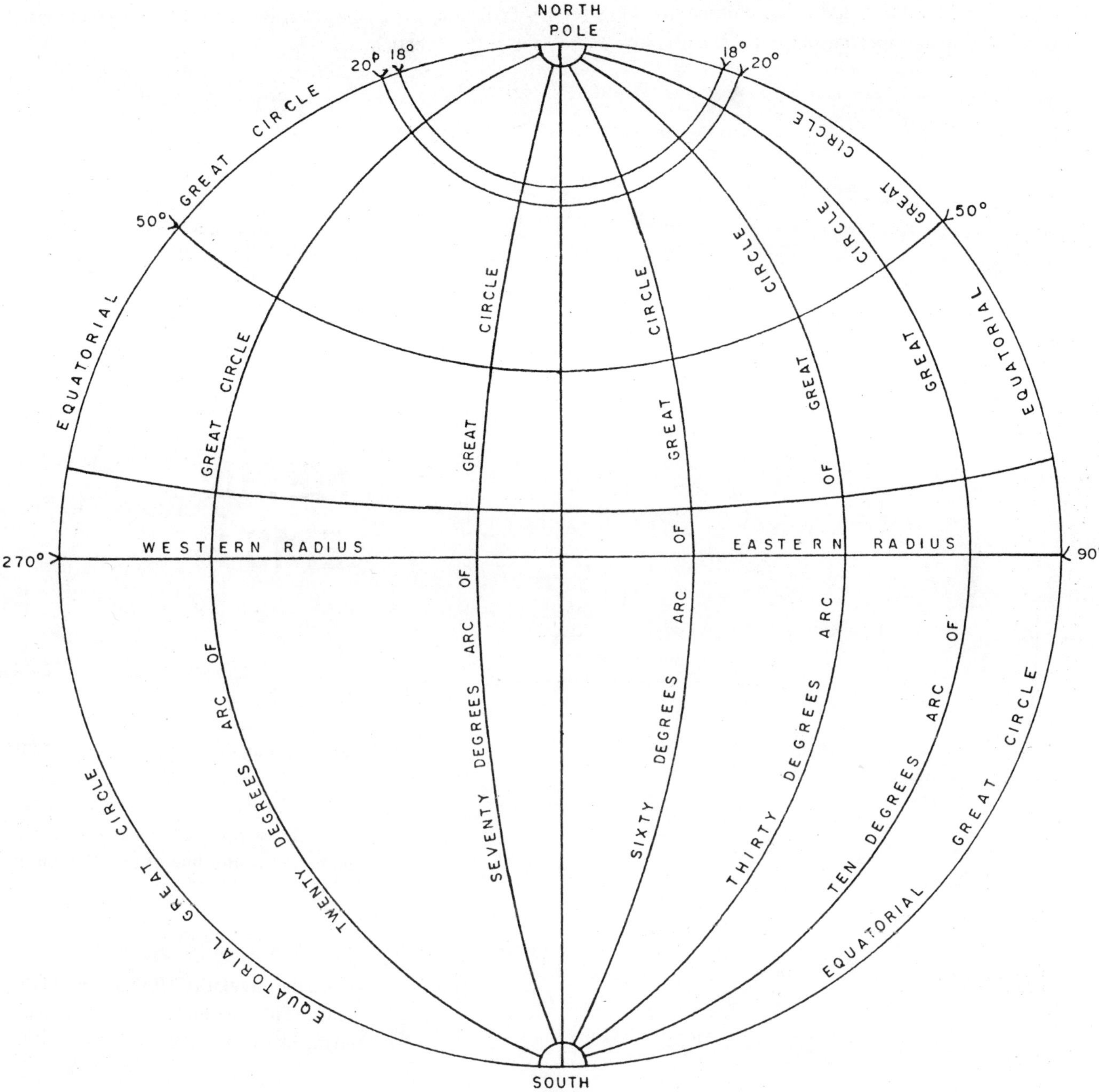

Fig. 5.1.0

Fig. 5.1.0 **STEREOGRAPHICNET (WULFES NET). SOME PARTS ARE SHOWN**

Equitorial great circle = Primitive circle = Circumference. Arcs of great circle = longitude like lines joining north and south poles.

East and West radii = used for marking inclination of planes.

The latitudinal lines = used for marking inclination of planes.

Note that the curvature of either longitude or latitude changes.

The longitude has maximum curvature near the circuference while the latitude has maximum curvature at the poles.

N 40 °E, N 60 °E, S 80 °E, S 20 °W, N 40 °W and so on (Fig. 5.1.1.). So in order to plot the given direction, the proper meridian is to be chosen and its meeting place with the primitive circle, locates the direction.

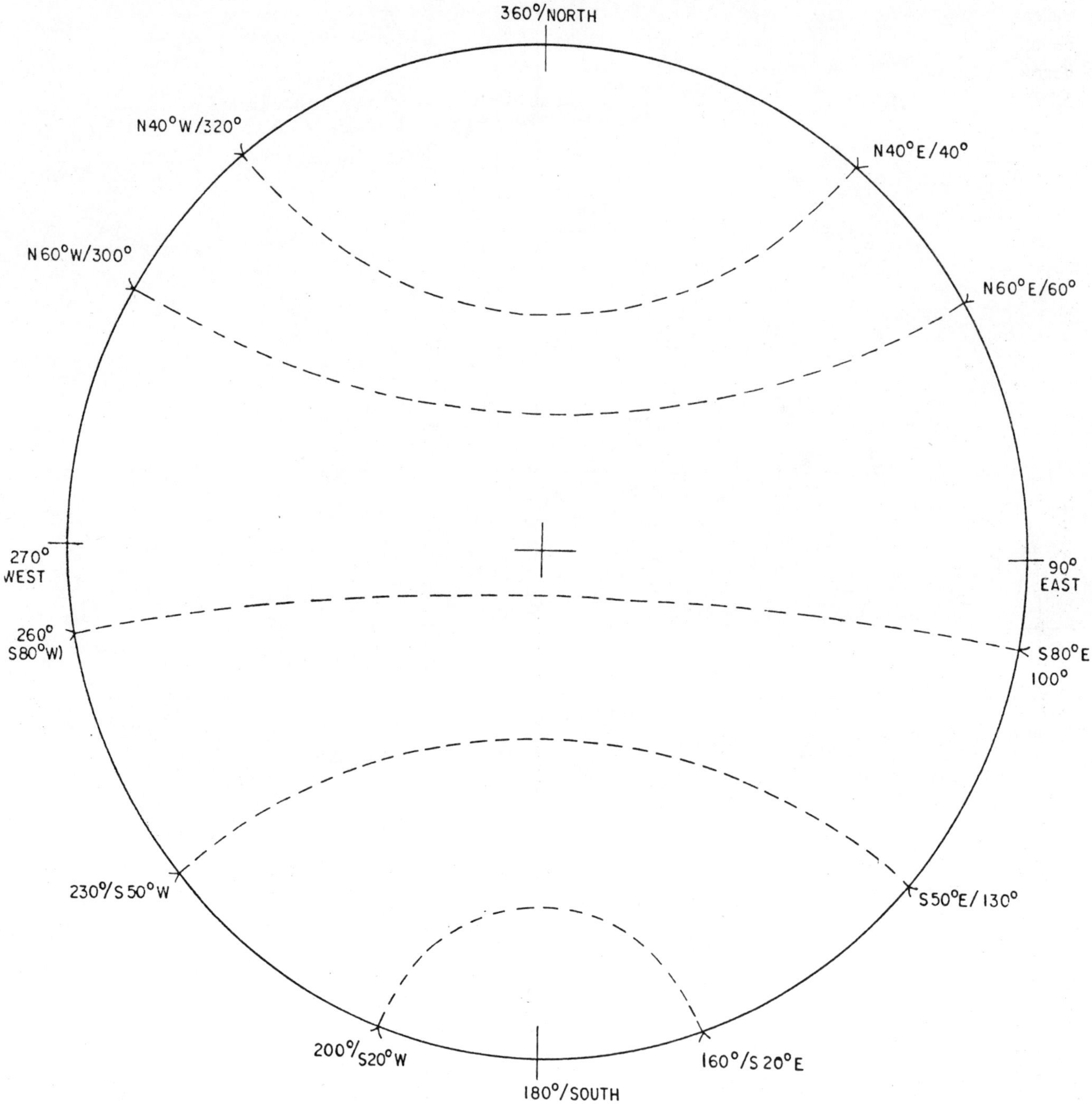

Fig. 5.1.1

Fig. 5.1.1 Shows the Positions of Some of the Cardinal, Circular/Quadrantal Bearings

NOTE : These are fixed by the (latitude) arcs of great circle touching the primitive or equitorial great circle. Such points when joined to the centre of the net, give the directions of the lines.

5.1.3 Role of arcs of Great Circles

These are used to mark the inclination of a plane or a surface. The point of intersection of any arc of great circle with the eastern or the western radius, marks the angle of inclination. At no other place the inclination of the given plane is to be marked. The periphery meaning the primitive circle, stands for zero degrees and the centre of the net stands for 90 degrees inclination. The intermediate values like 5, 30, 50, 70 etc., take places in between 0° and the centre of the net (Fig 5.1.2).

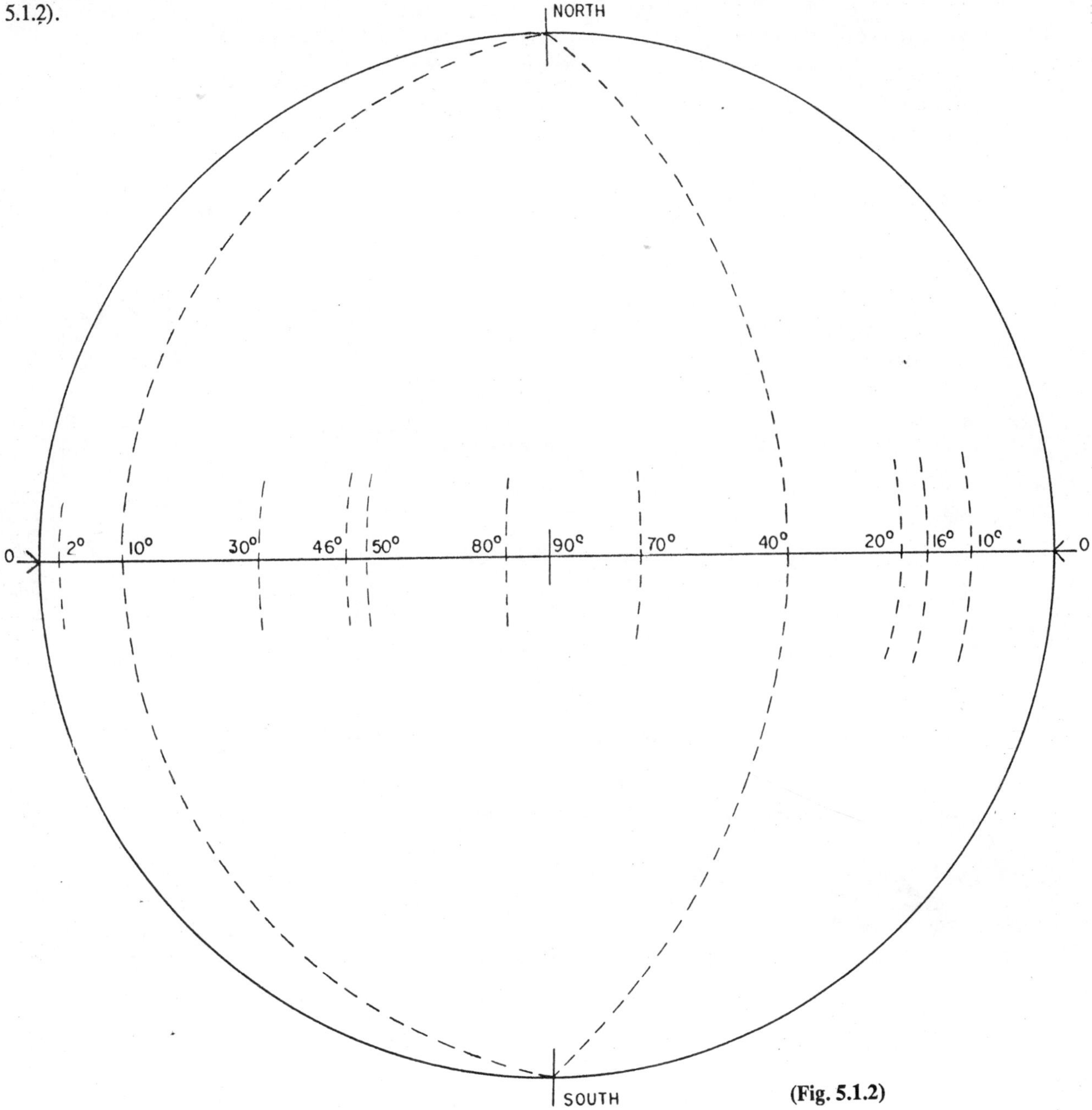

(Fig. 5.1.2)

Fig. 5.1.2 Shows Inclined Arcs of Great Circles (Longitudes)

NOTE : Intersection of the longitude at the east or the west radius, yields the amount of inclination of the arc. It is zero at the primitive circle (easternmost and westernmost point) and 90 degrees at the centre. Note that the distance between consecutive arcs of great circle like 70°, 40°, 20°, 10°, etc., is not constant. It is widest near the circumference, and it is shorther near the centre of the net.

5.1.4 Procedure of Plotting Dipping Surface

Problem 5.1 Represent a plane having a dip of 60° due S 60 °W.

Procedure

In stereographic projection, the direction of dip and the amount are most important. Merely from the strike direction, the actual directioin of dip cannot be fixed, and hence the plotting cannot be achieved. Also it is not necessary to give the direction of strike, as it can be derived from that of the direction of dip, because these two directions are perpendicular to each other.

Place the tracing sheet over the stereographic net, and draw the equatorial geat circle (the primitive great circle). Indicate the centre of the circle by a cross. Draw north line (Fig. 5.1.3) half inside, half outside of the circumference of the

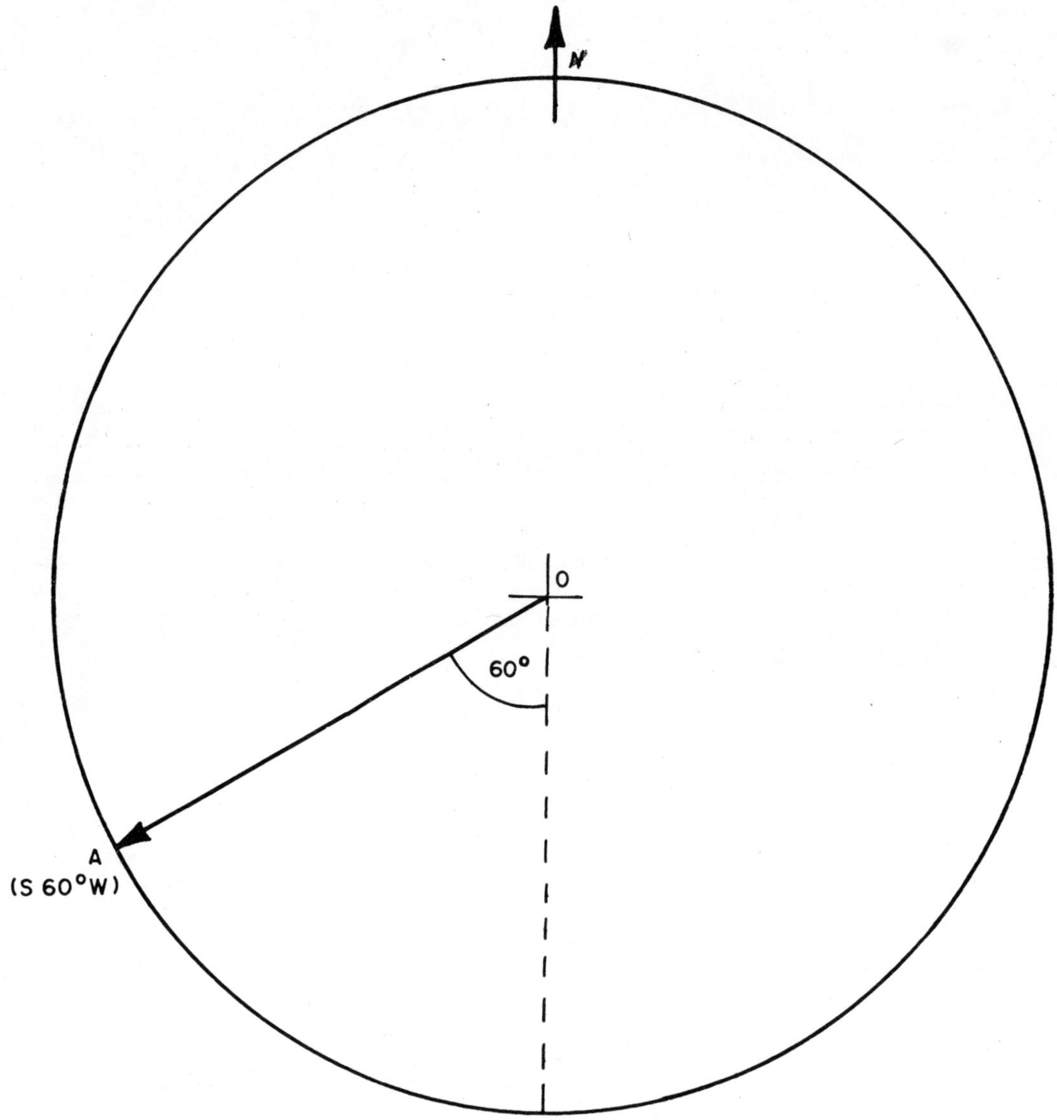

Fig. 5.1.3

Fig. 5.1.3 Marketing the Position of Dip Direction *OA;* due S 60 °W on the Tracing Paper (Problem 5.1)

circle. No other lines are to be drawn like, the east, the west etc. Mark off 40° from south towards west and obtain point A. Join *OA* and place an arrow at the end of the line *OA*, but it should be inside the circle (Fig. 5.1.3). Rotate the tracing sheet clockwise and make *OA* to coincide with the western radius as seen through the tracing sheet. In this position obtain point *A'* on *AO*, such that it is 20° away from the periphery. Draw an arc of great circle through the point A upto the Poles - *D* and *D'*. Draw line *DOD'*(Fig. 5.1.4). Bring the tracing sheet to zero position i.e., the north line drawn on the tracing sheet should coincide with the north line of the stereographic net (Fig. 5.1.5). Check up the sterogram i.e., verify whether *OA* is due S 60°W and that *DOD'* gives the trend of N 30 °W - S 30 °E.

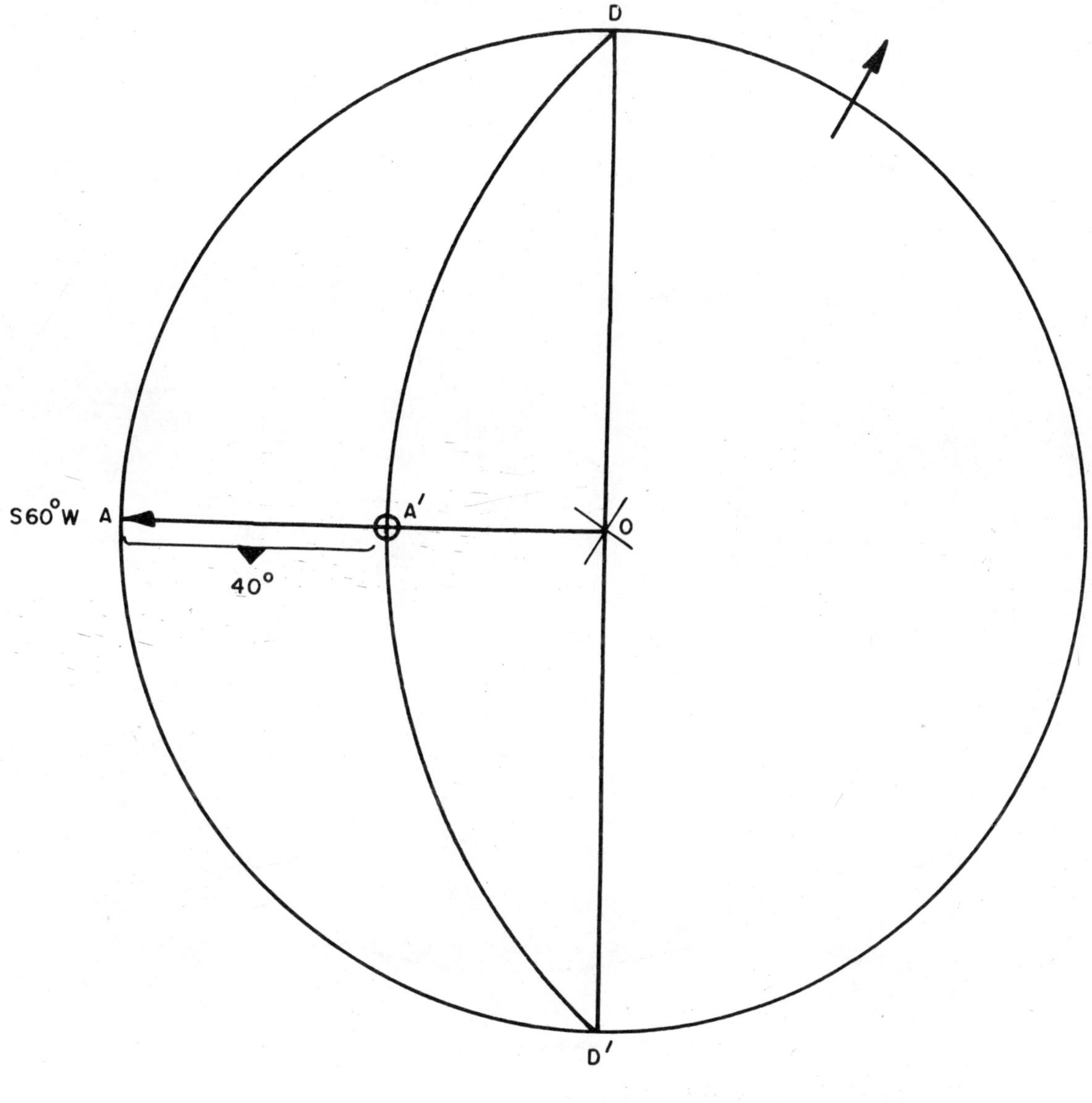

Fig. 5.1.4

Fig. 5.1.4 Rotating the tracing paper and bringing *OA* to lie over western radius of the net. Point *A'* and the arc of great circle through it, are drawn in this position. *DOD'* is also drawn in this position (Problem 5.1)

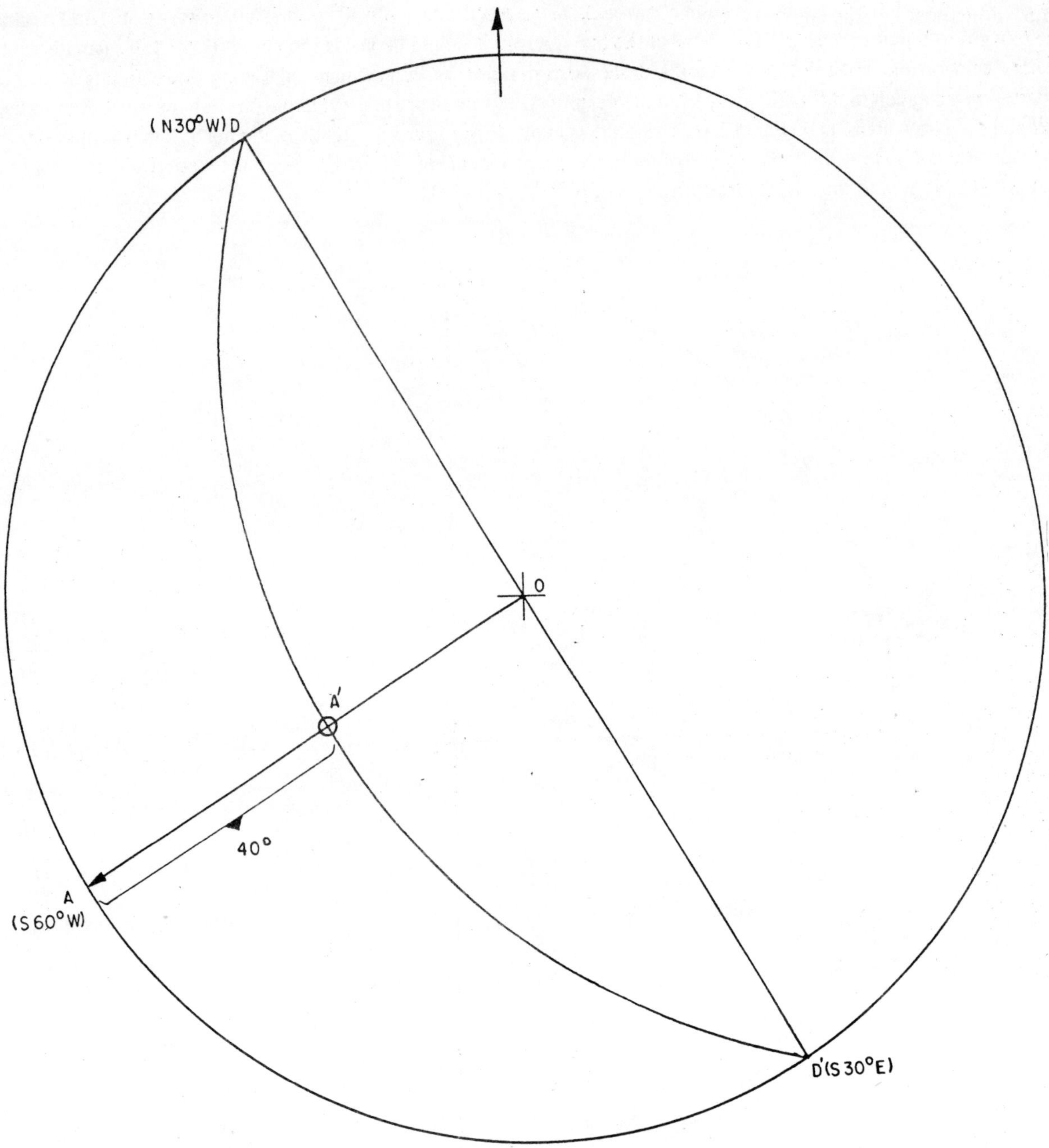

Fig. 5.1.5

Fig. 5.1.5 Plane Dipping 40° Due S 60 °W (Problem 5.1)

NOTE : The tracing paper is in zero position i.e., the north of the tracing paper has coincided with the north of the net. OA = S 60 °W = true dip direction

AA'= 40°, DOD'= true strike direction = N 30 °W – S30 °E

5.1.5 Functions of the Different Parts of a Stereogram

In Fig. 5.1.5 a plane dipping 40° due S 60 °W has been shown. But it is necessary to indicate as to which part constitutes the dipping plane, which line controls the values of apparent dip and so on. Therefore these features have been indicated in Fig. 5.1.6

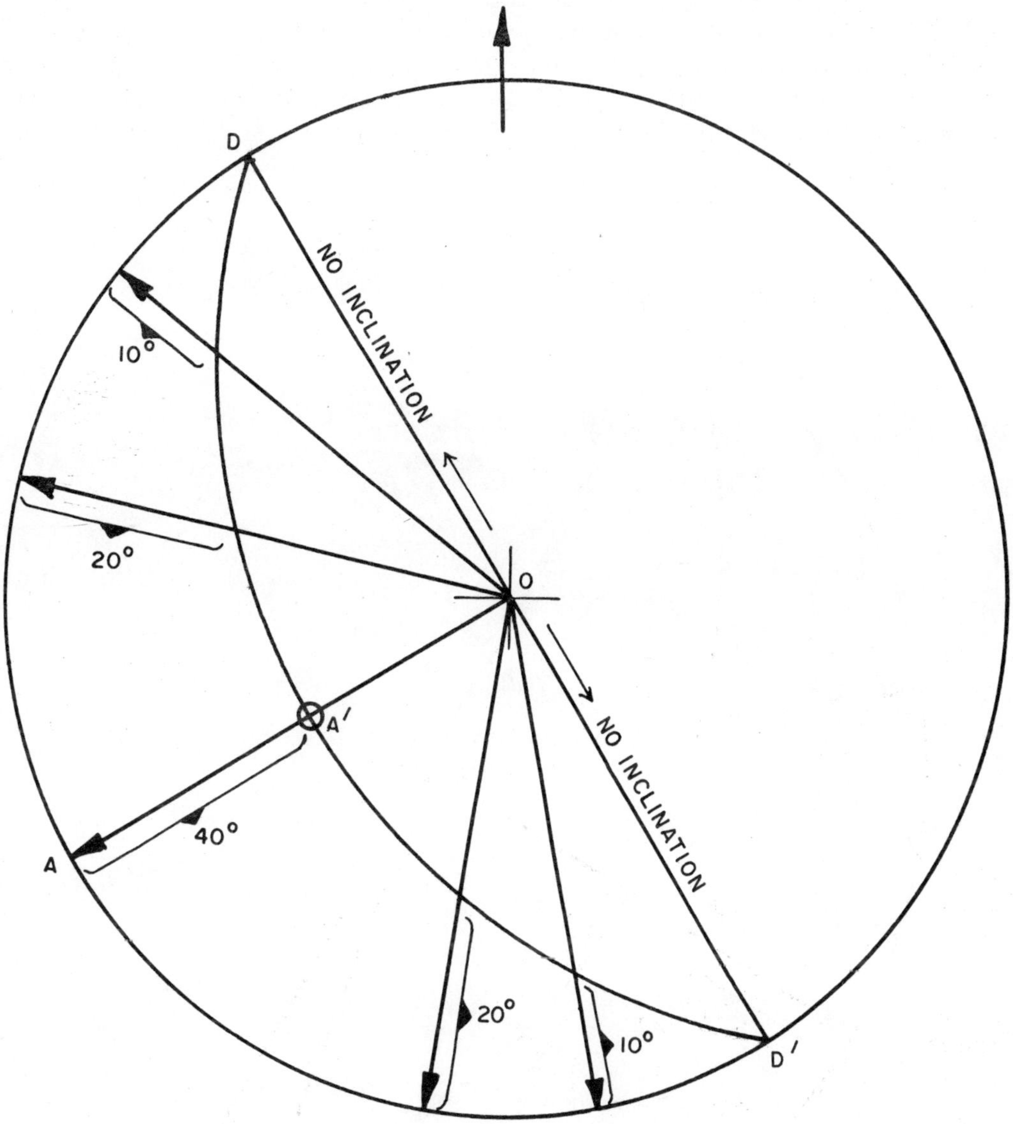

Fig. 5.1.6.

Fig. 5.1.6 Parts of Dipping Plane (Problem 5.1)

stippled part = dipping plane *DOD'* = strike line.

DA'D= dipping surface, *DAD'* = reference circle or primitive circle.

Note that the (angular) distance *DA'D'* and *DAD'* decreases from *AA'* towards *D* and *D'* ends. At D and D there is no distance between the two surfaces, and therefore the inclination in direction *OD* or *OD'*, is zero. Also note that the inclination of the plane decreases on both sides of line OA. This means that there will be two directions along which identical amount of apparent dip like 20°, 10°, etc., are possible. These have been shown in the diagram.

In Fig. 5.1.6 the stippled area stands for the dipping plane and the distance between the primitive circle *DAD'* and the dipping surface *DA'D'* measured along the concerned radius (this stands for the direction of the apparent dip) is a measure of the amount of inclination. It is seen clearly in Fig. 5.1.6 that this distance between *DAD'* and *DA'D'* is maximum between points *A* and *A'*, and it goes on becoming less on moving towards *D* and *D'* from *A'*. At *D* or *D'*, there is no separation between

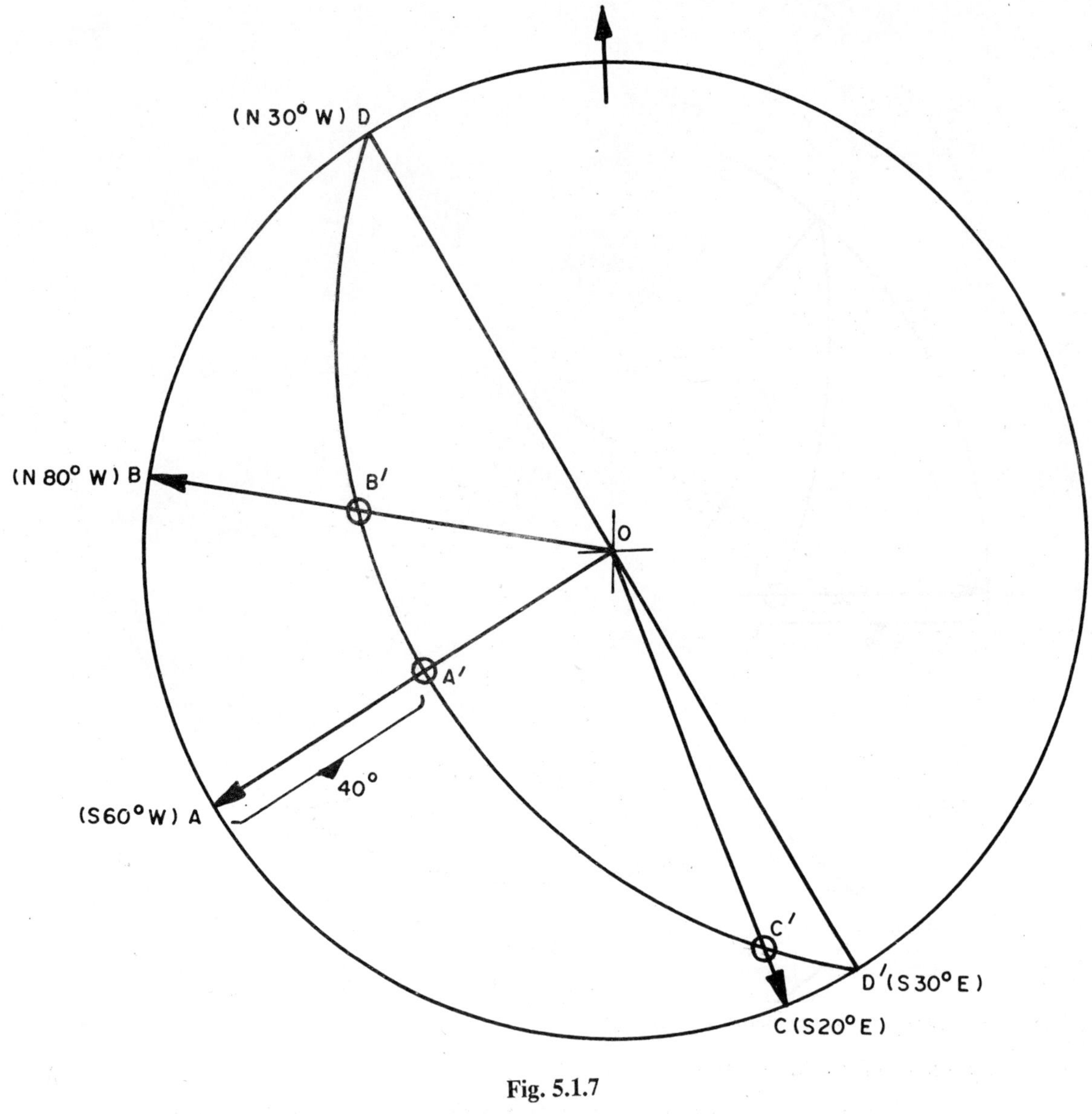

Fig. 5.1.7

Fig. 5.1.7. Plotting of Plane Dipping 40° due S 60 °W (as done Through Figs. 5.1.3, 5.1.4 and 5.1.5), and Apparent Dip Direction *OB* and *OC* due N 80 °W and S 20 °E (Problem 5.2)

Note that *OB* and *OC* have cut the arc of great circle *DA'D'* at *B'* and *C'*, but angle *BB'* or *CC'* is not measured in this position of the stereogram.

the primitive circle and the dipping plane. This situation in other words means that there is no inclination along *OD* or *OD'*. Therefore the line *DOD'* stands for the strike line.

Yet another feature deserves a mention. On either sides of *OA*, the amount of apparent dip goes on decreasing. Therefore there will be two directions of apparent dip along which the amount will be indentical. Thus in Fig. 5.1.6, there are two directions having a dip of 20°, two directions having a dip of 10° and so on. Thus it is to be remembered that there are two directions along which identical amount of apparent dip is obtainable. This is not possible for the true dip, unless it is a fold.

5.1.6 Procedure of Reading Apparent Dip in a Desired Direction

Problem 5.2 A formation has a true dip of 40° due S 60 °W. Find out the amount of apparent dip due N 80 °W and S 20 °E.

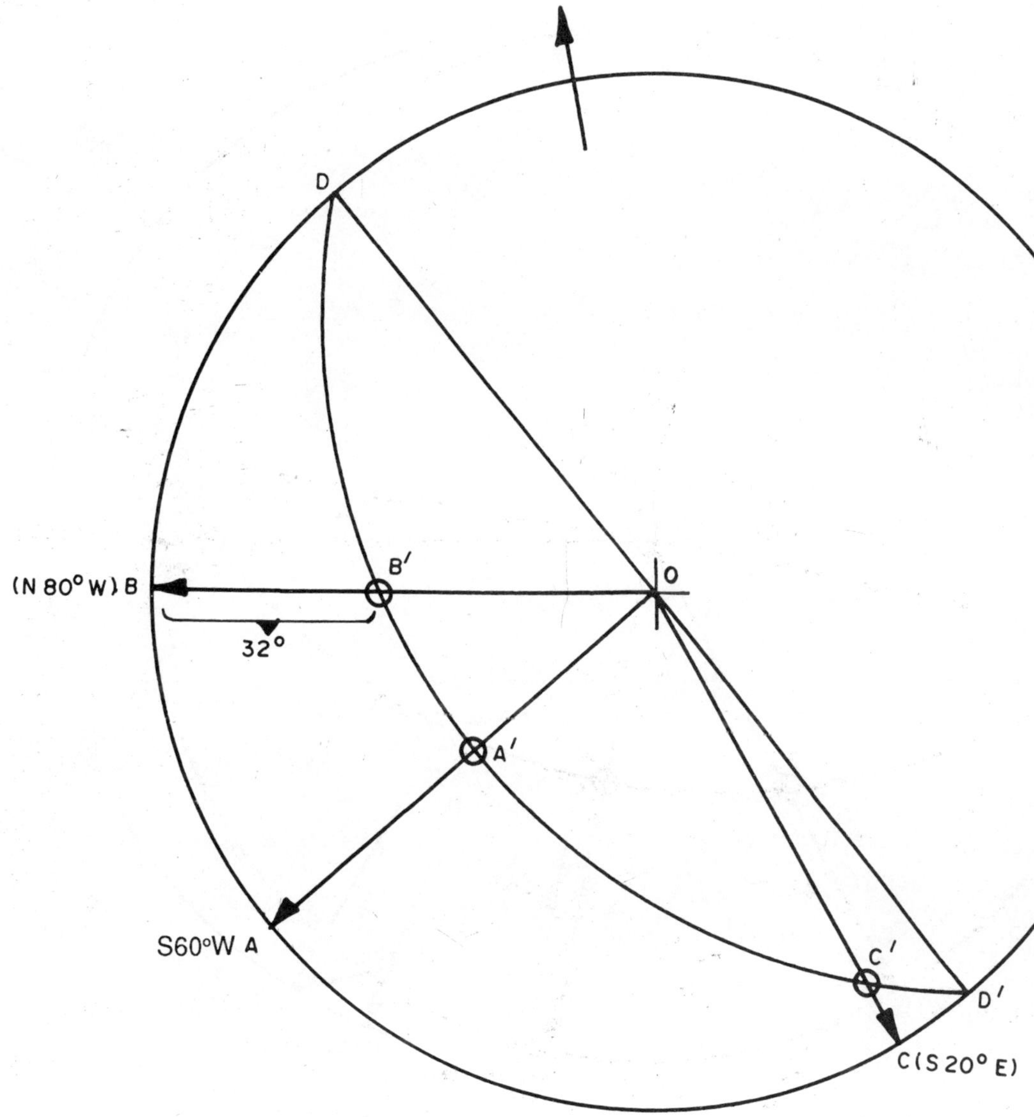

Fig. 5.1.8

Fig. 5.1.8 Bring *OB'B* to Coincide with Western Radius and Measure Angle *BB'* which is found to be 32° (Problem 5.2)

Procedure

Following the procedure laid down in 5.1.4 above, draw a sterogram to represent the plane dipping 40° due S 60 °W (Fig. 5.1.7). In the zero position of the stereogram draw lines *OB* and *OC* due N 80 °W and S 20 °E respectively (Fig. 5.1.7). These lines intersect the dipping surface *DA'D'* at *B'* and *C'*, respectively. Rotate the tracing sheet and bring *OBB'* to coincide with the western radius of the stereographic net and read off the angle *BB'* (Fig. 5.1.8). Likewise bring *OC'C* to coincide with the eastern radius of the stereographic net and read off angle *CC'* (Fig. 5.1.9). The amounts of dip due N 80°W (*BB"*) and S 20 °E (*CC'*) are found to be S 32°E and 8° respectively. Finally bring the stereogram to zero position (Fig. 5.2.0).

5.1.7 Procedure of Obtaining Apparent Dip Direction having Desired Amount of Dip

Problem 5.3 A plane dips 40° due S 60 °W. In which directions, the amount of dip will be 10°?

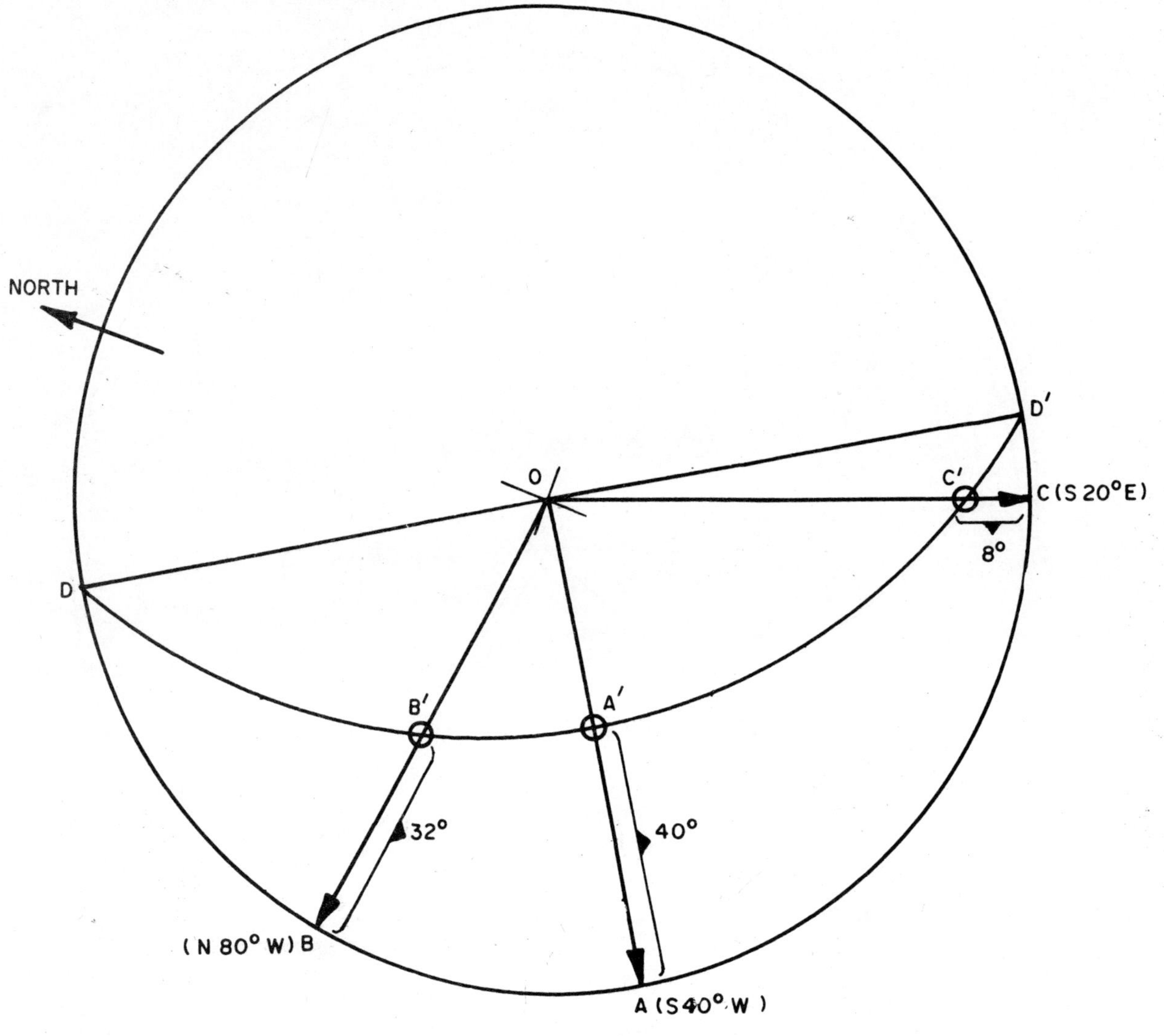

Fig. 5.1.9

Fig. 5.1.9 Bring *OC'C* to Coincide with Eastern Radius and Measure Angle *CC'* which is found to be 8° (Problem 5.2)

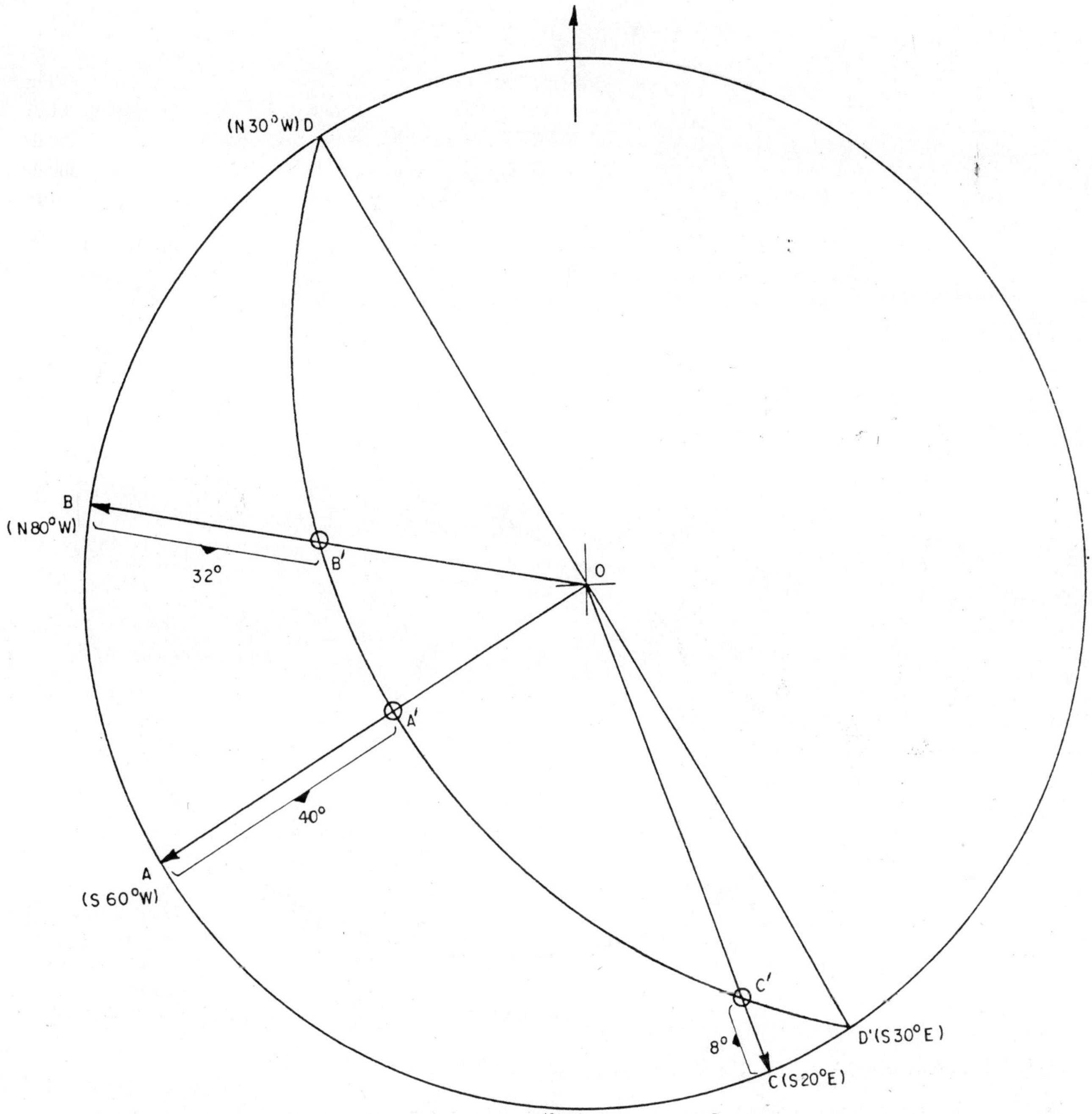

Fig. 5.2.0

Fig. 5.2.0 Derivation of Apparent Amount Due N 80 °W and S 20 °E, given True Dip of 40° due S 60 °W (Problem 5.2)

NOTE *BB'* and *CC'* are 32° and 8° respectively, but these angles are read off in Figs 5.1.8 and 5.1.9 respectively. The tracing paper is in zero position.

Procedure

Following procedure laid down in 5.1.4 above, draw a stereogram to represent the plane dipping 40° due S 60 °W (Fig. 5.2.1).. Rotate the tracing sheet (clockwise or anticlockwise) and watch the points of intersection made by the dipping

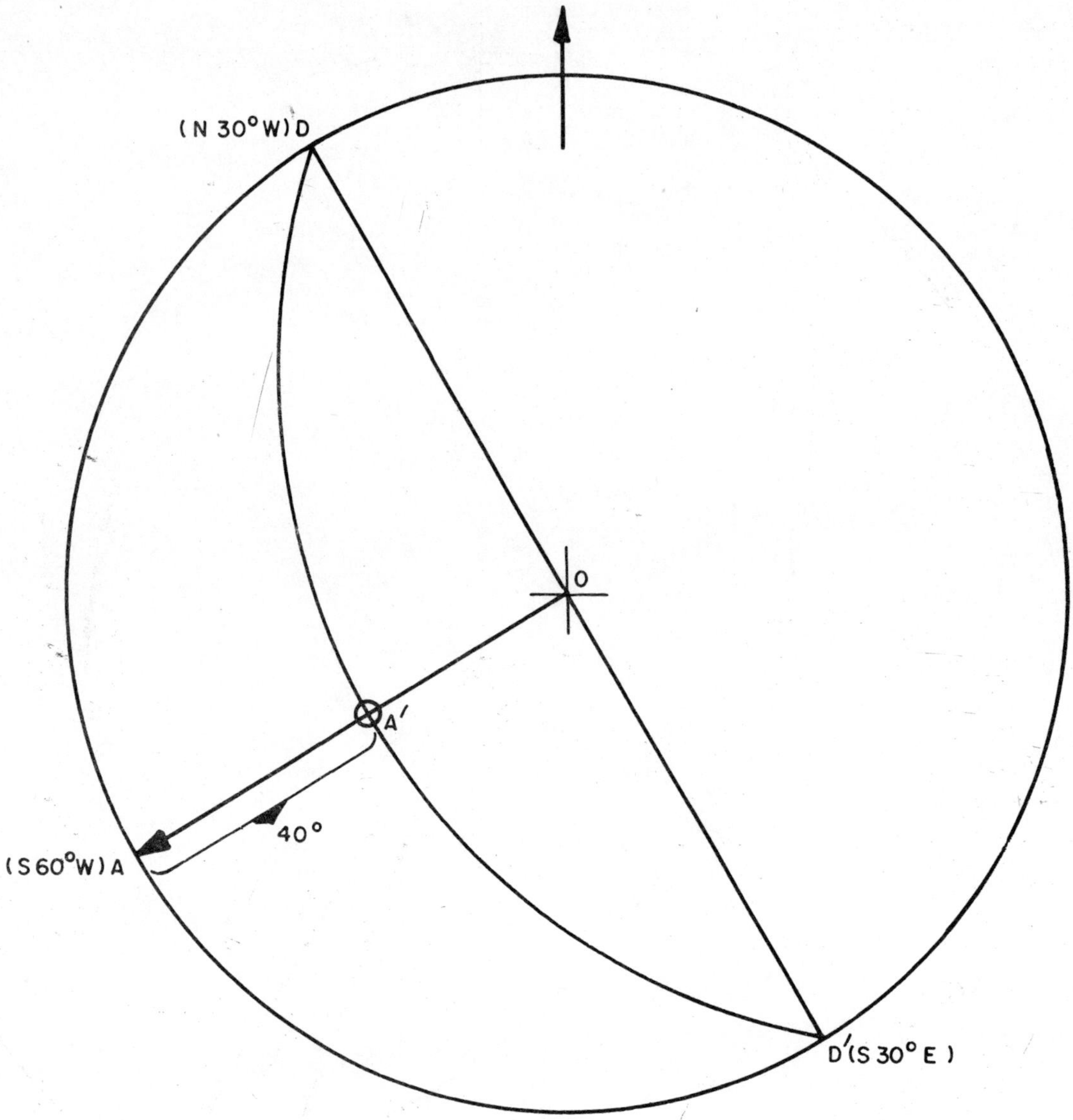

Fig. 5.2.1

Fig. 5.2.1 A Plane Dipping 40° due S 60 °W (Problem 5.3)

NOTE The stereogram is in zero position, but it is derived by following steps given in Figs. 5.1.3, 5.1.4 and 5.1.5.

surface *(DA'D')* on the western or eastern radius. Keep on rotating the tracing sheet till a position is obtained where the amount of dip is 10°. This has been shown in Fig. 5.2.2. Trace the western radius (because the dipping surface *DA'D'* has intersected the western radius), i.e., draw the line *OB'*,B on the tracing sheet. This gives the direction along which the amount of dip is 10° *(BB')*, *but the actual direction is not to be read off in this position (Fig. 5.2.2) because the stereogram is not in zero position.*

Keep on rotating the tracing sheet further, till another position is obtained where the dipping surface *DA'D'* intersects the eastern radius giving a dip of 10°. This has been shown in Fig. 5.2.3. Draw the eastern radius (because in this position,

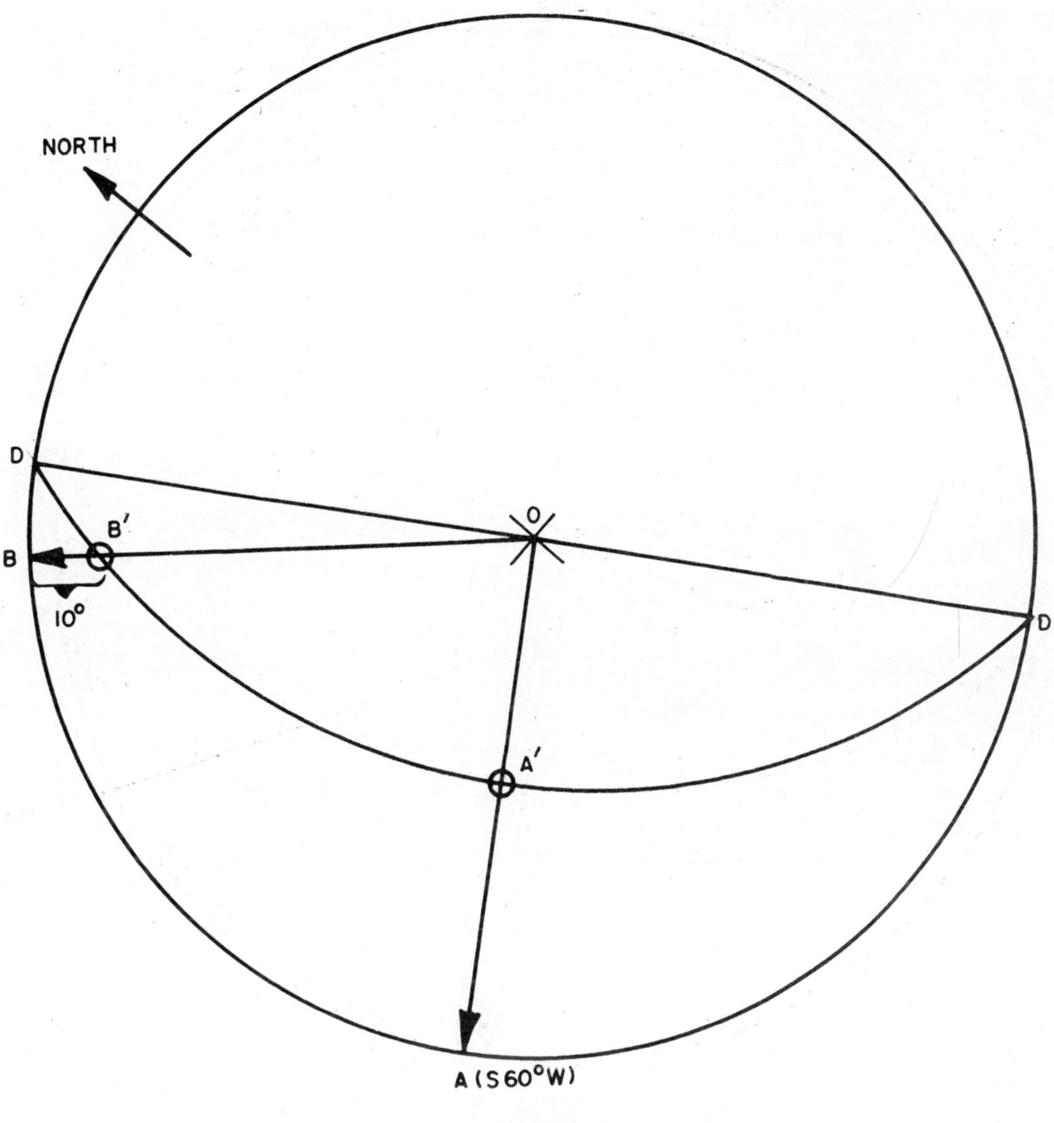

Fig. 5.2.2

Fig. 5.2.2 Rotate the tracing paper and make the dipping surface *DA′D′* to cut the western radius at point *B′* such that *BB′* is of value of 10° (Problem 5.3)

Note that *OB′B* is along the western radius of the net as seen below the tracing paper.

the dipping surface *DA′D′* has intersected the eastern radius), i.e, draw the line *OC′C*. This gives the second direction along which the amount of dip against is 10° *(CC′). This however (Fig. 5.2.3) does not give the actual direction of dip because the stereogram is not in zero position.*

Now bring the stereogram to zero position (Fig. 5.2.4) and read off directions off *OB′B* and *OC′C*. These are found to be N 43 °W and S 18 °E.

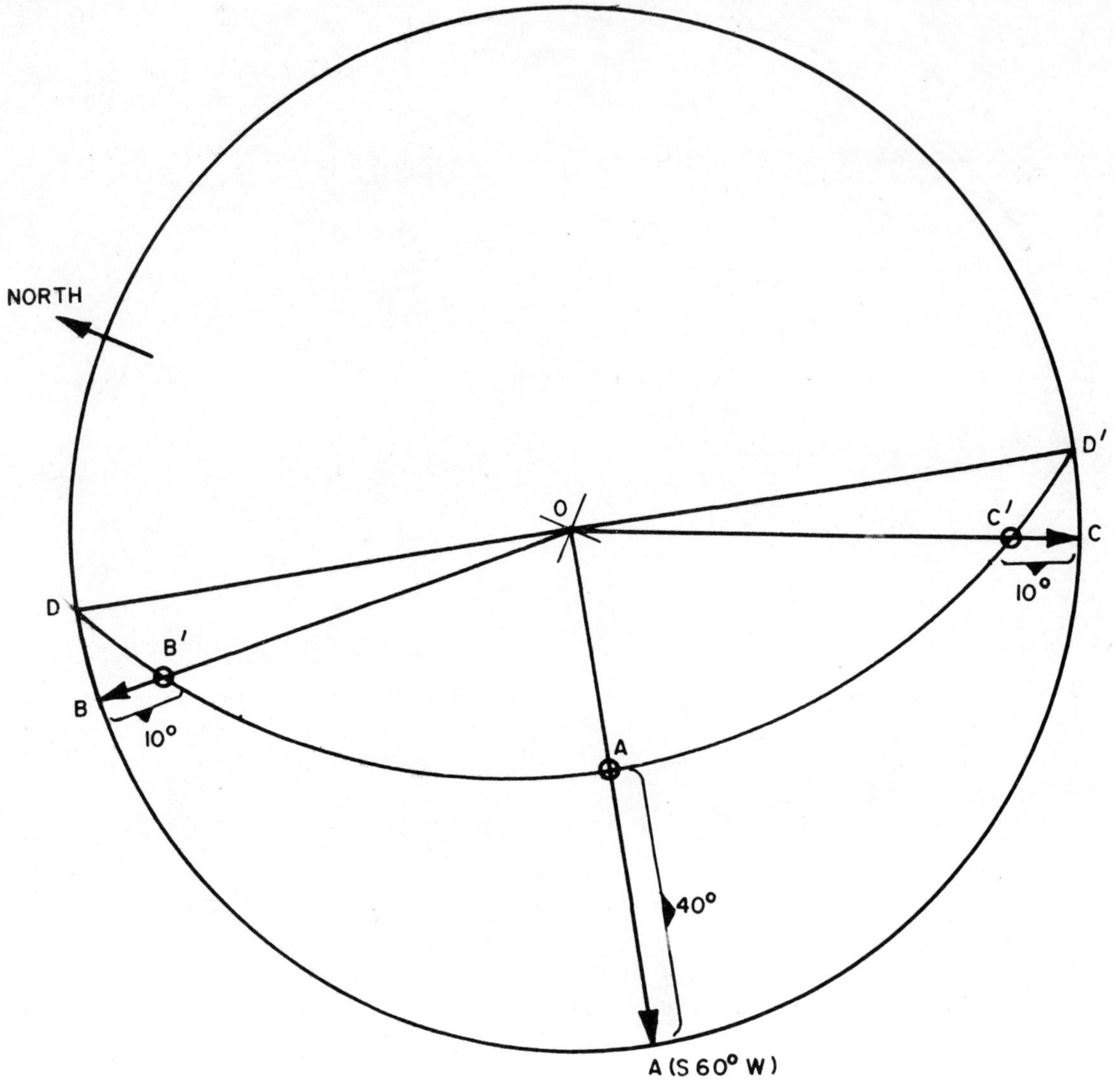

Fig. 5.2.3

Fig. 5.2.3 Rotate the tracing paper and now make the dipping surface $DA'D'$ **cut the Eastern radius at point** C' **having a value of 10° (Problem 5.3) Note that** $OC'C$ **marks the eastern radius of the net as seen below the tracing paper.**

5.1.8 Determination of True Dip from two Apparent Dips

There are several ways in which the data can be set like giving the two apparent dips, one apparent dip and true dip direction, one apparent dip and true strike and so on. The procedure of obtaining the true dip amount or direction will be described in following pages.

(a) True Dip from two Apparent Dips

Problem 5.4 A formation has an apparent dip of 20° due N 40 °W and 50° due N 60 °E. Find the direction of true dip, amount of true dip and direction of true strike.

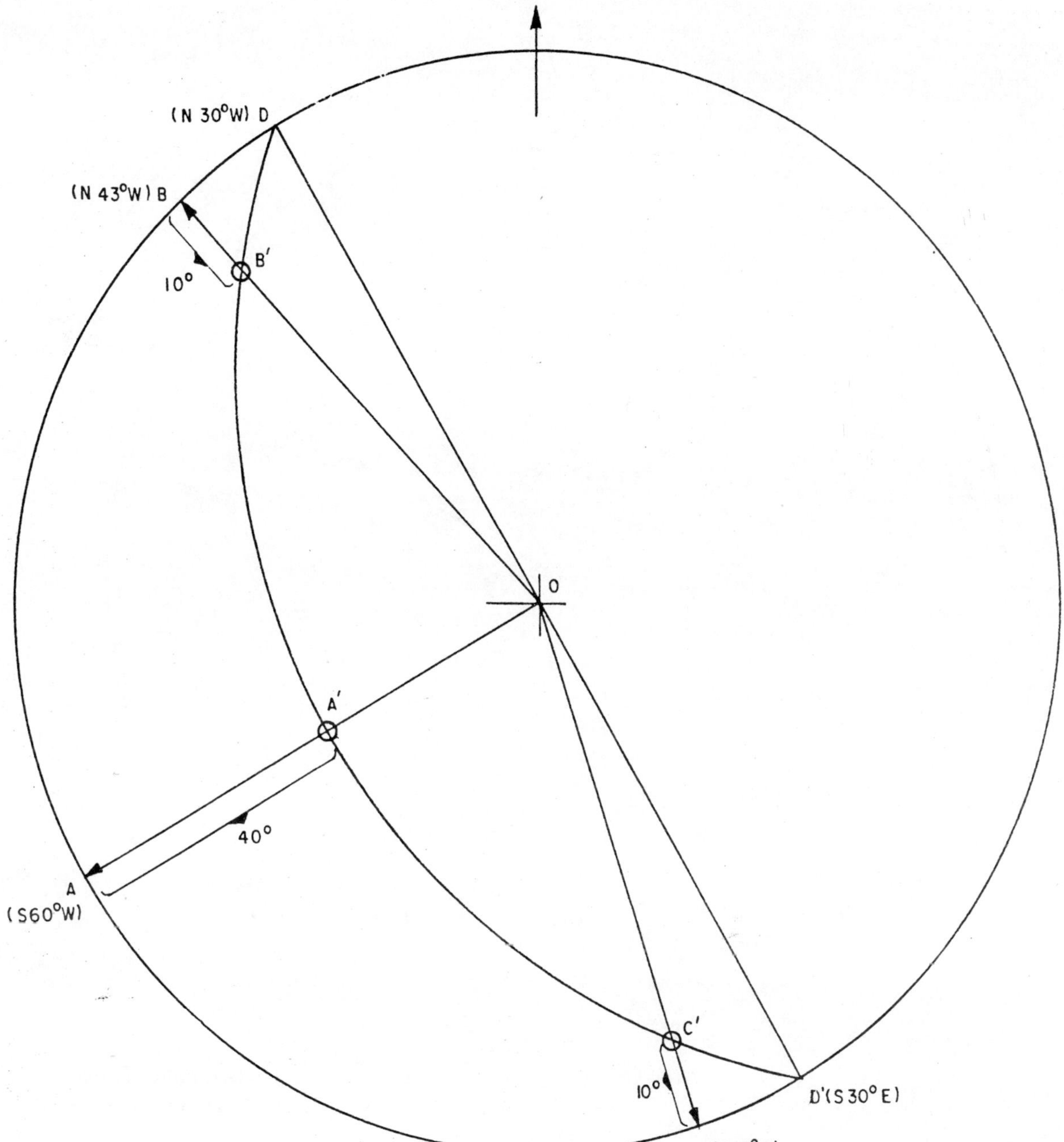

Fig. 5.2.4

Fig. 5.2.4. Derivation of Apparent Dip Direction having Amount of Dip of 10° (Problem 5.3)

NOTE : *OB'B* and *OC'C* are the two directions of apparent dip which are due N 43 °W and S 18 °E, respectively. These bearings are read only in the zero position of the stereogram.

Procedure

Trace the equitorial great circle (circumference), mark the centre and the north line. Draw lines *OA* and *OB* due N 40 °W and N 60 °E respectively as shown in Fig. 5.2.5. Rotate the tracing sheet and bring *OA* to coincide with the western radius and mark off point A' such that AA' is equal to 20° (Fig. 5.2.6). Bring OB to coincide with the eastern radius and mark off point B' such that BB' is equal to 50° (Fig. 5.2.7).

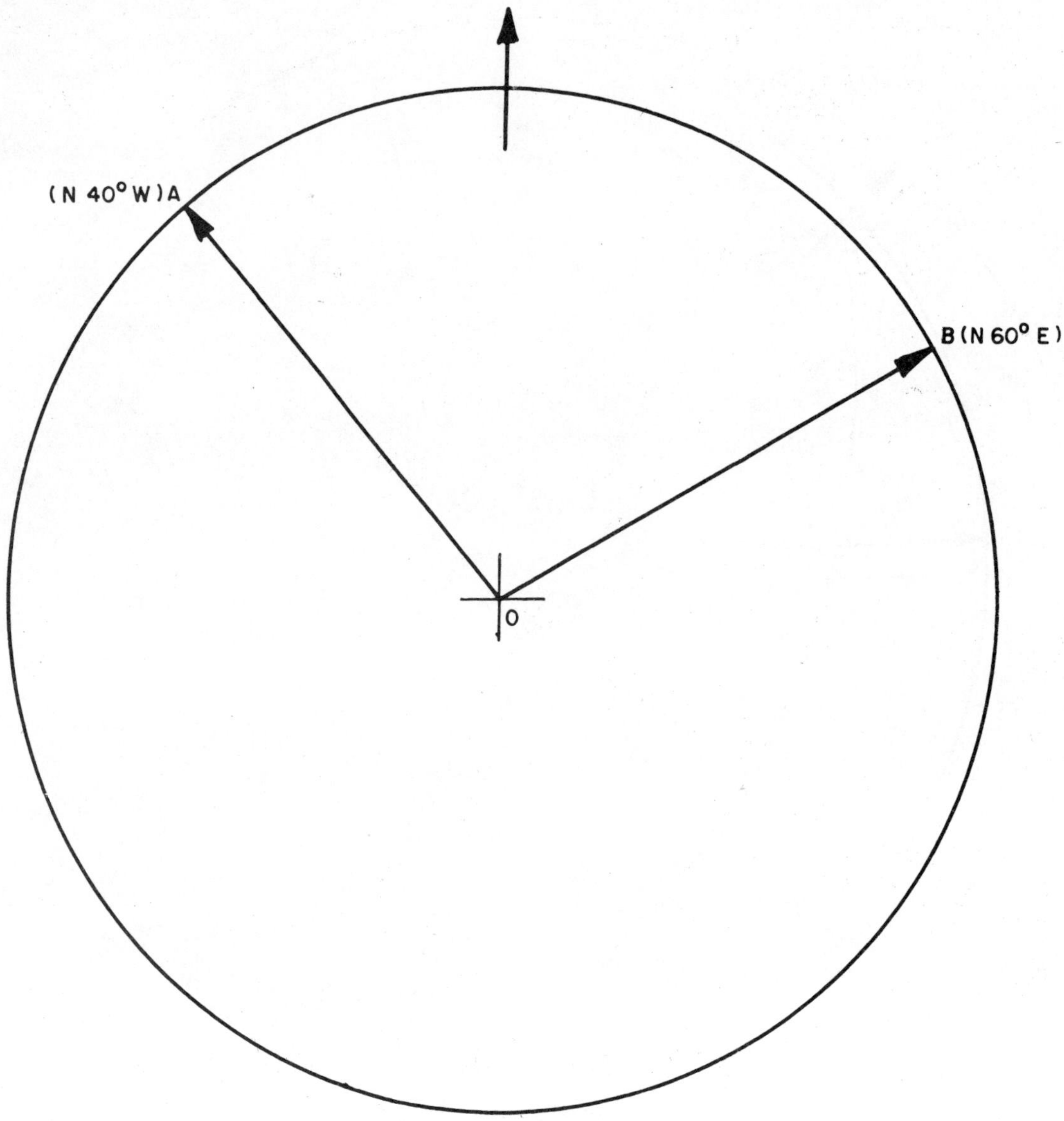

Fig. 5.2.5

Fig. 5.2.5 Plotting Apparent Dip Directions Due N 40 °W (OA) and N 60 °E (OB) (Problem 5.4)

Now rotate the tracing sheet till points A' and B' fall on the same arc of circle. Points A' and B' may be taken on the eastern side or the western side of the stereographic net (Figs. 5.2.8 and 5.2.9). When both A' and B' have fallen on the same arc of circle of the net, trace that arc upto the northern and the southern poles of the stereographic net (Fig. 5.2.8 and 5.2.9). Now without disturbing the position of the paper draw the eastern radius (if A' and B' have fallen on the eastern side of the net as in Fig. 5.2.8) or the western radius (if the points A' and B' have fallen on the western side of the net, as in Fig. 5.2.9). Also draw a line joining the northern and the southern poles, which passes through the centre of the stereogram (line *DOD'* in Figs 5.2.8 and 5.2.9). The lines drawn along the eastern and western radius in Figs. 5.2.8 and 5.2.9 have been labelled as *OC'C*.

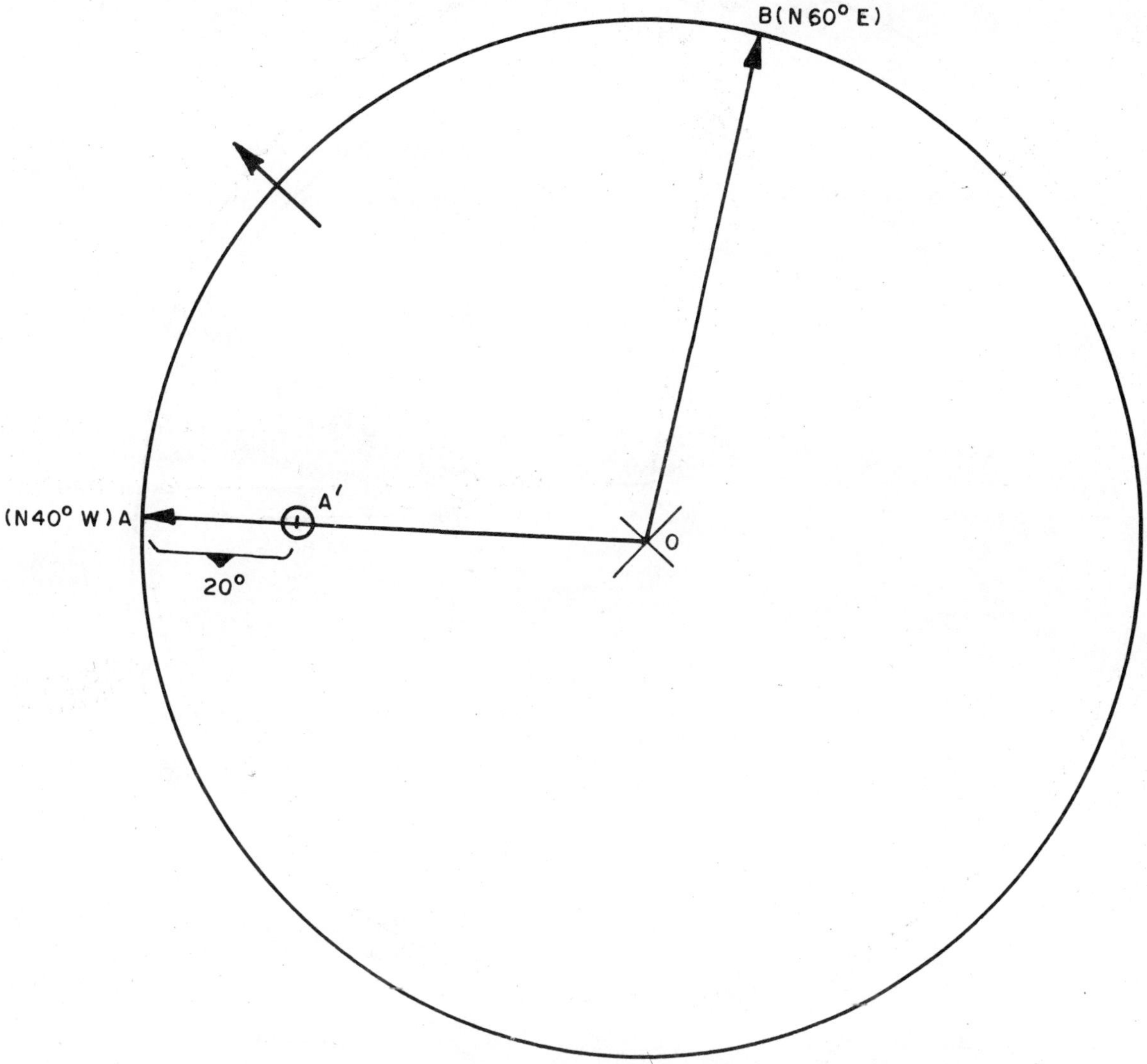

Fig. 5.2.6

Fig. 5.2.6 Plotting Apparent Dip Amount of 20° along *OA* (Problem 5.4)

NOTE :*AA'* is the amount of dip of 20°, which is plotted after making *OA* to coincide with western radius.

After drawing the arc of great circle connecting points *A* and *B*, the line *DOD* and line *OCC*, the angle *CC* is measured which gives the amount of true dip. This is found to be 52° (Figs. 5.2.8 and 5.2.9). However in this position where the amount of dip is measured, neither the direction of true strike (line *DOD*) nor the direction of true dip (*OCC*) are to be measured, *because the stereogram is not in zero position.* Therefore rotate the stereogram to zero position, and now measure the direction *OCC* and *DOD* which are the directions of true dip and true strike, respectively (Fig. 5.3.0). It is found to be N 35 °E and N 55 °W – S 55 °E respectively.

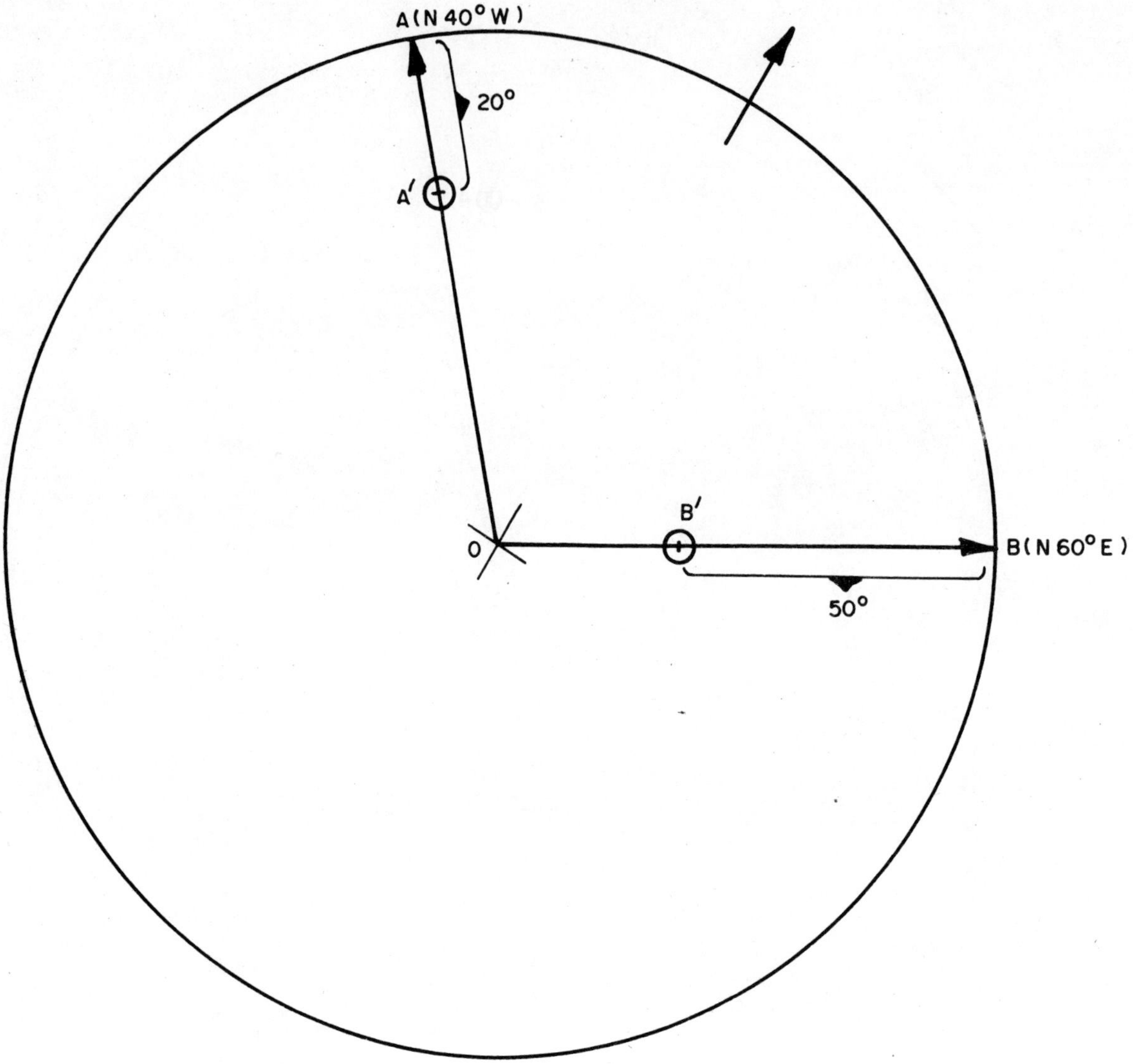

Fig. 5.2.7

Fig. 5.2.7 Plotting Apparent Dip Amount of 50° along OB (Position 5.4)

NOTE : *BB′* is the amount of dip of 50° which is plotted after making *OB* to coincide with the eastern radius.

(b) One Apparent Dip and Direction of True Dip

Problem 5.5. A formation has an apparent dip of 30° due N 40 °E. If the true dip direction be due S 70 °E, find the amount of true dip and the direction of true strike.

Procedure

Draw line *OA* due N 40 °E and a line *OB* due S 70 °E as shown in Fig. 5.3.1. Now rotate the tracing sheet and make line *OA* to coincide with the eastern radius and mark off 30° and obtain point *A′* (Fig. 5.3.2). Rotate the tracing sheet and bring line *OB* to coincide with the eastern radius (it can be made to coincide with the western radius also, the final result does

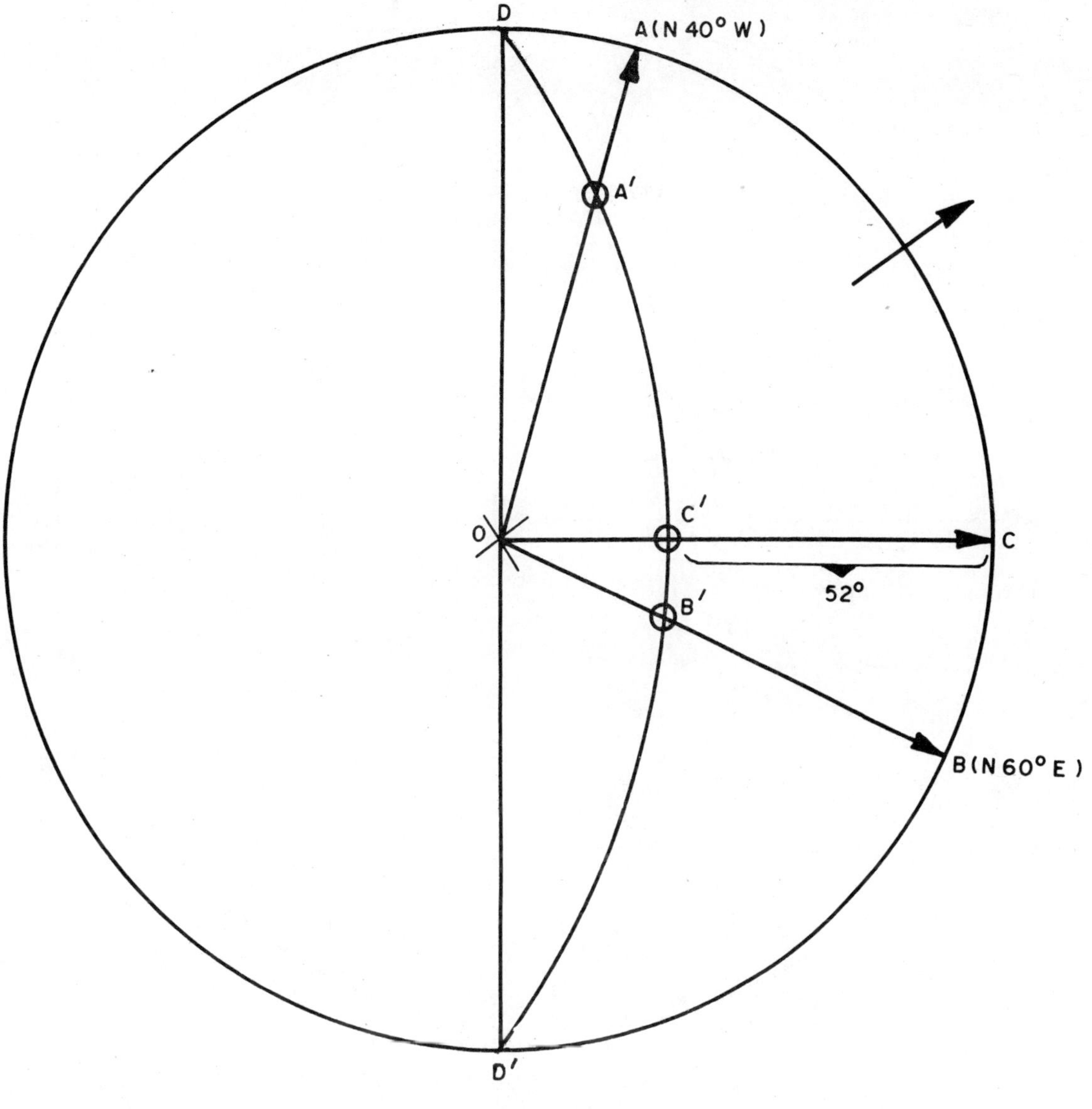

Fig. 5.2.8

Fig. 5.2.8 Disposition of Tracing Paper when A' **and** B' **Fall on the Same Arc of Great Circle. But the Points Fall on the Eastern Part of the Net (Problem 5.4)**

NOTE : A' and B' have fallen on the eastern side of the net. Trace eastern radius (line OC) which cuts the arc of great circle, through $A'B'$ at point C'. $OC'C$ is the direction of true dip, but it is to be read off only when the stereogram is rotated to zero position (Fig. 5.3.0). Read off angle CC' which gives the amount of true dip as 52°.

not change at all), and now draw that arc of the great circle on which the point A' has fallen (Fig. 5.3.3). The arc of great circle will interesect the line OB at B (Fig. 5.3.3). Measure BB' which gives the amount of true dip (59°, Fig. 5.3.3). Draw the N – S diameter (line DOD' in Fig. 5.3.3) which stands for the true strike line. Now rotate the tracing sheet and bring it to zero position. (Fig. 5.3.4). Read the position taken by the line $OB'B$ and DOD' which stand for the direction of true dip and true strike. It is found that the true strike direction is N 20 °E – S 20 °W and the amount of true dip is 59° (Fig. 5.3.3).

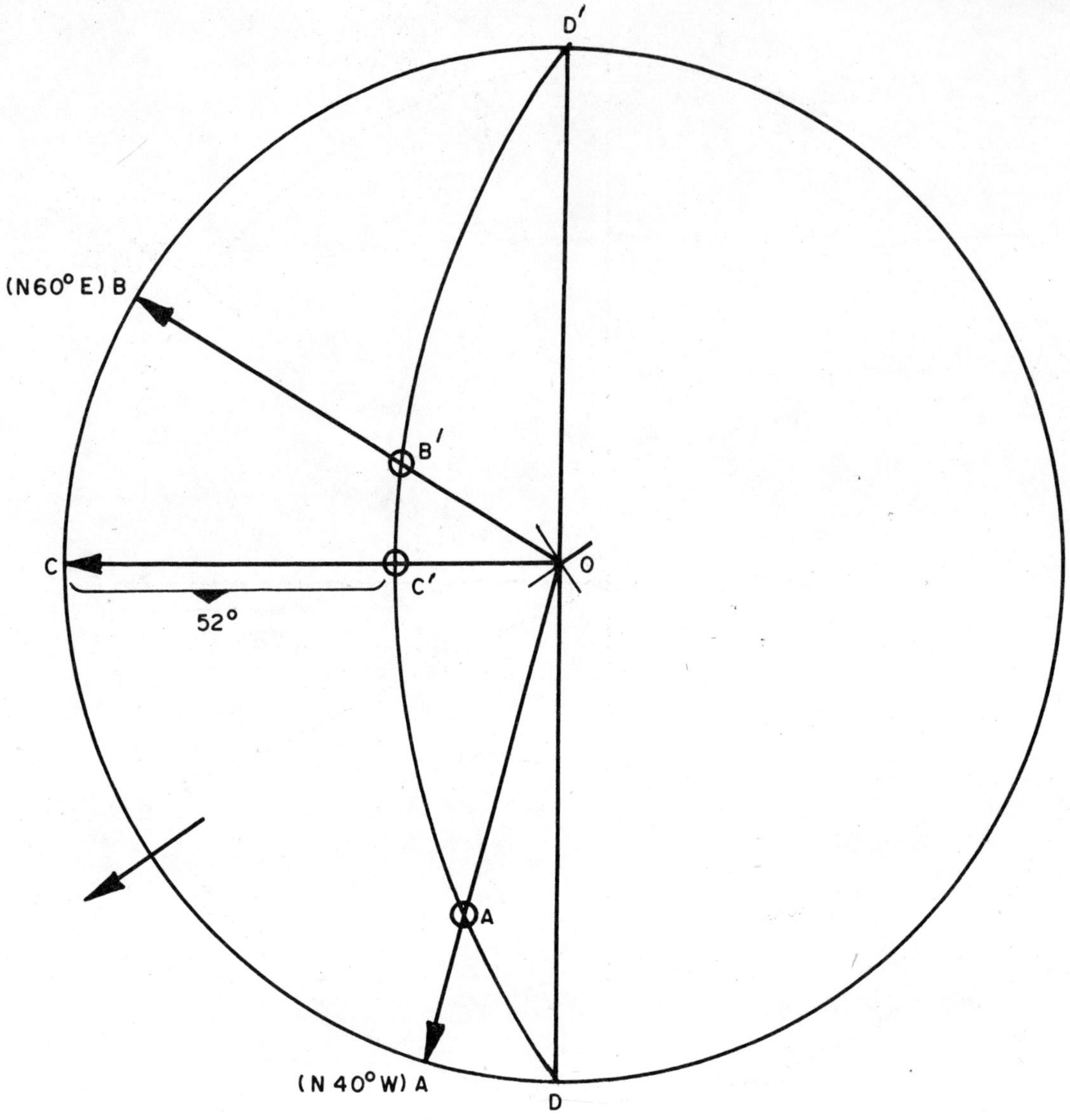

Fig. 5.2.9

Fig. 5.2.9 Disposition of Tracing Paper When *A′B′* Fall on Same Arc of Great Circle, but the Points Fall on the Western Part of the Net (Problem 5.4)

NOTE : *A′B′* have fallen on the western part of the net. Trace western radius (line *OC*) which cuts the arc of great circle through *A′B′* at point *C′*. *OC′C* is the direction of true dip, but it is to be read off only when the tracing paper is brought to zero position (Fig. 5.3.0). Read off angle *CC′* which gives the amount of true dip as 52°.

(c) One Apparent Dip and True Strike Direction

Problem 5.6 A formation has an apparent dip of 40° due east. If the true strike direction be N 50 °E-S 50 °W, find the direction and amount of true dip.

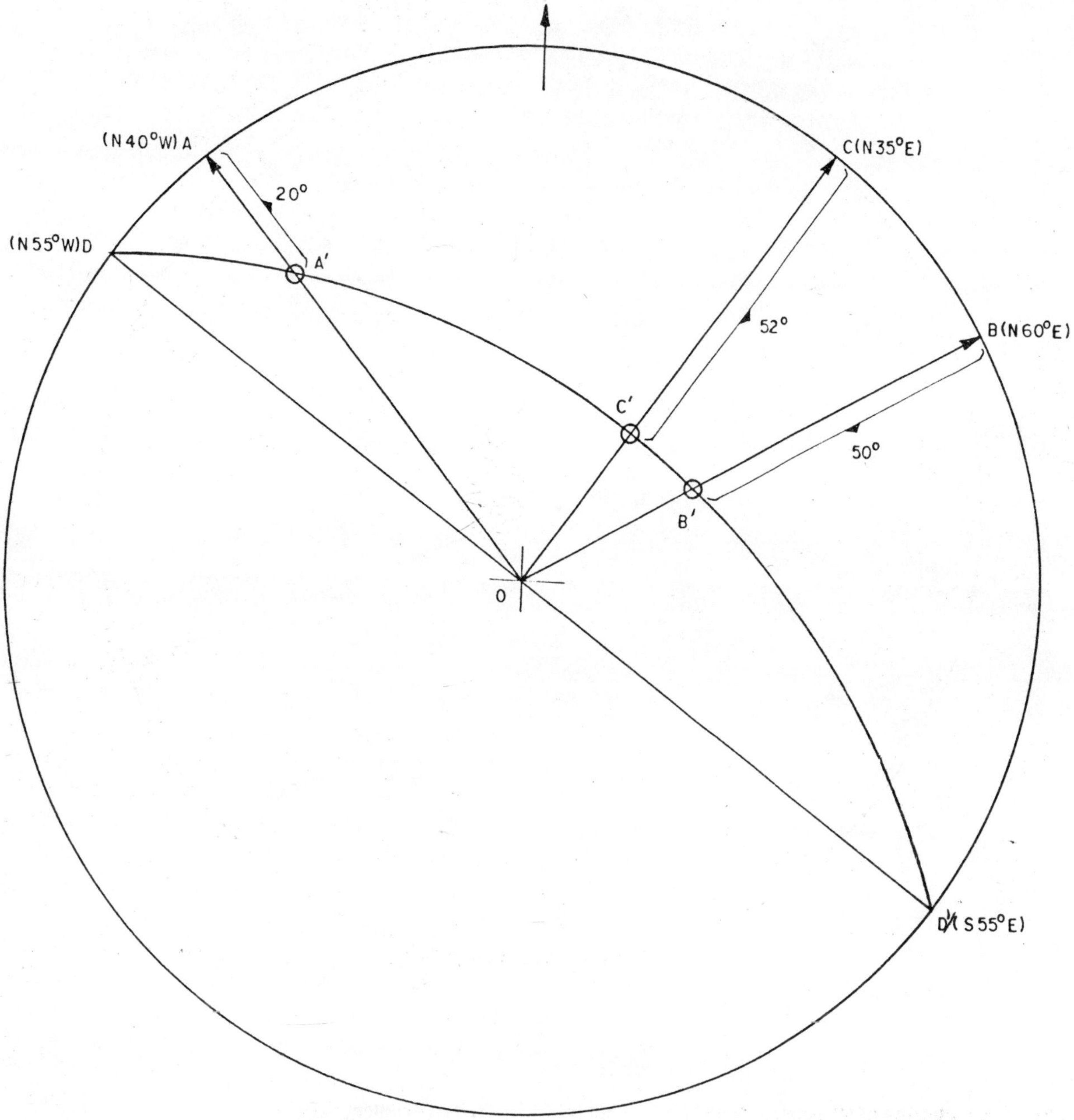

Fig. 5.3.0

Fig. 5.3.0 Derivation of True Dip Direction and Amount from Two Apparent Dip Directions (Problem 5.4)

NOTE : $OC'C$ = true dip direction = N 35 °E

DOD' = true strike direction = N 55 °W– S 55 °E

CC' = amount of true dip = 52° (This is to be measured either in Fig. 5.2.8 or 5.2.9).

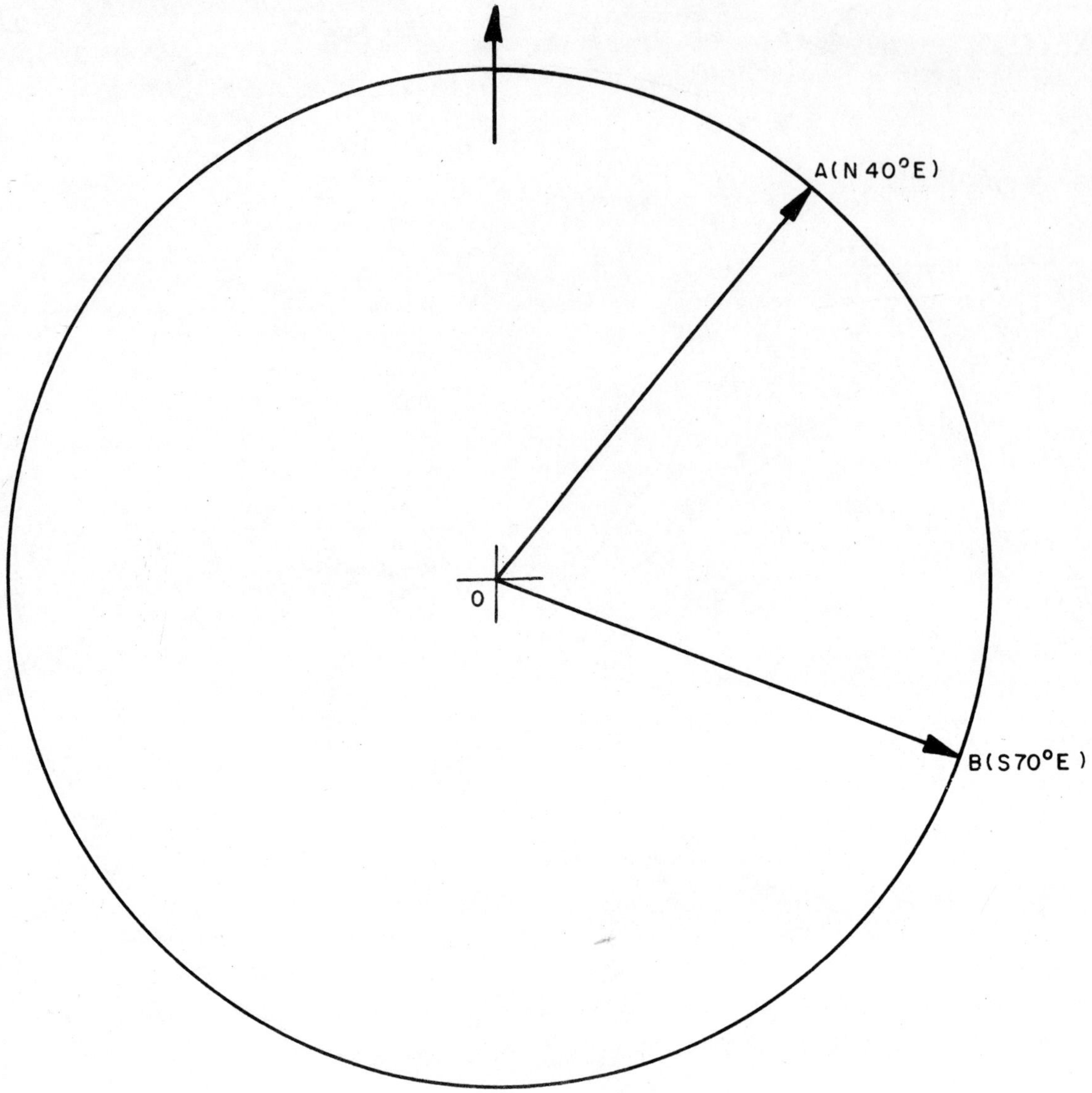

Fig. 5.3.1

Fig. 5.3.1 Plotting of Directions N 40 °E (OA) and S 70 °E (OB) (Problem 5.5)

NOTE :· Only directions are to be drawn and not the amount of inclination along them.

Procedure

In earlier problem (Problem 5.5.), the direction of true strike was derivable without carrying out the actual plotting of the data. But in the present case, though the direction of true strike is given, yet the actual direction of true dip cannot be derived with certainty, without plotting of the data. In problem 5.6, the dip could be due N 40 °W or due S 40 °E. The actual direction can be derived only through plotting of the given direction of apparent dip.

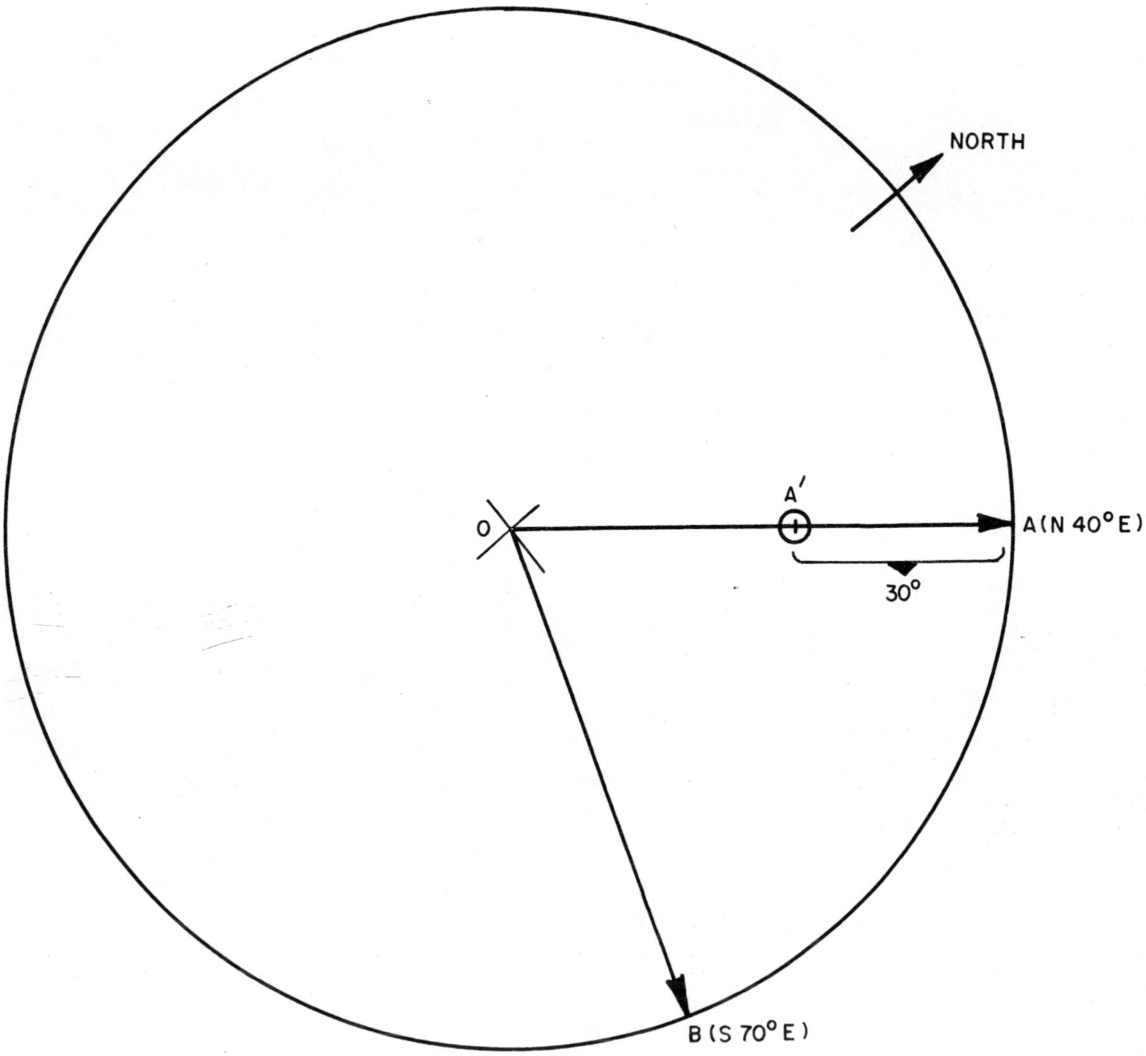

Fig. 5.3.2

Fig. 5.3.2 Rotating *OA* to Coincide with Eastern Radius (Problem 5.5)

NOTE : In this position the amount of inclination of 30° (distance *AA*) is plotted and point *A'* obtained.

Draw *OA* due east and obtain points *A'* on it such that *AA'* is equal to 40° (Fig. 5.3.5). Also draw the given direction of true strike of N 50 °E – S 50 °W (line *DOD'* in Fig. 5.3.5). The apparent dip amount of 40° was plotable in the zero position of the stereogram, because the apparent dip direction coincided with the eastern radius of the net. Rotate the tracing sheet and make the line *DOD'* to coincide with the N–S diameter of the net (Fig. 5.3.6) and draw that arc of the great circle on which the point *A'* has fallen. Also trace the eastern radius (line *OB* in Fig. 5.3.6) which intersects the arc of great circle drawn through point *A'*, at point *B'*. Measure *BB'* which gives the amount of true dip (53° in Fig. 5.3.6). Rotate the

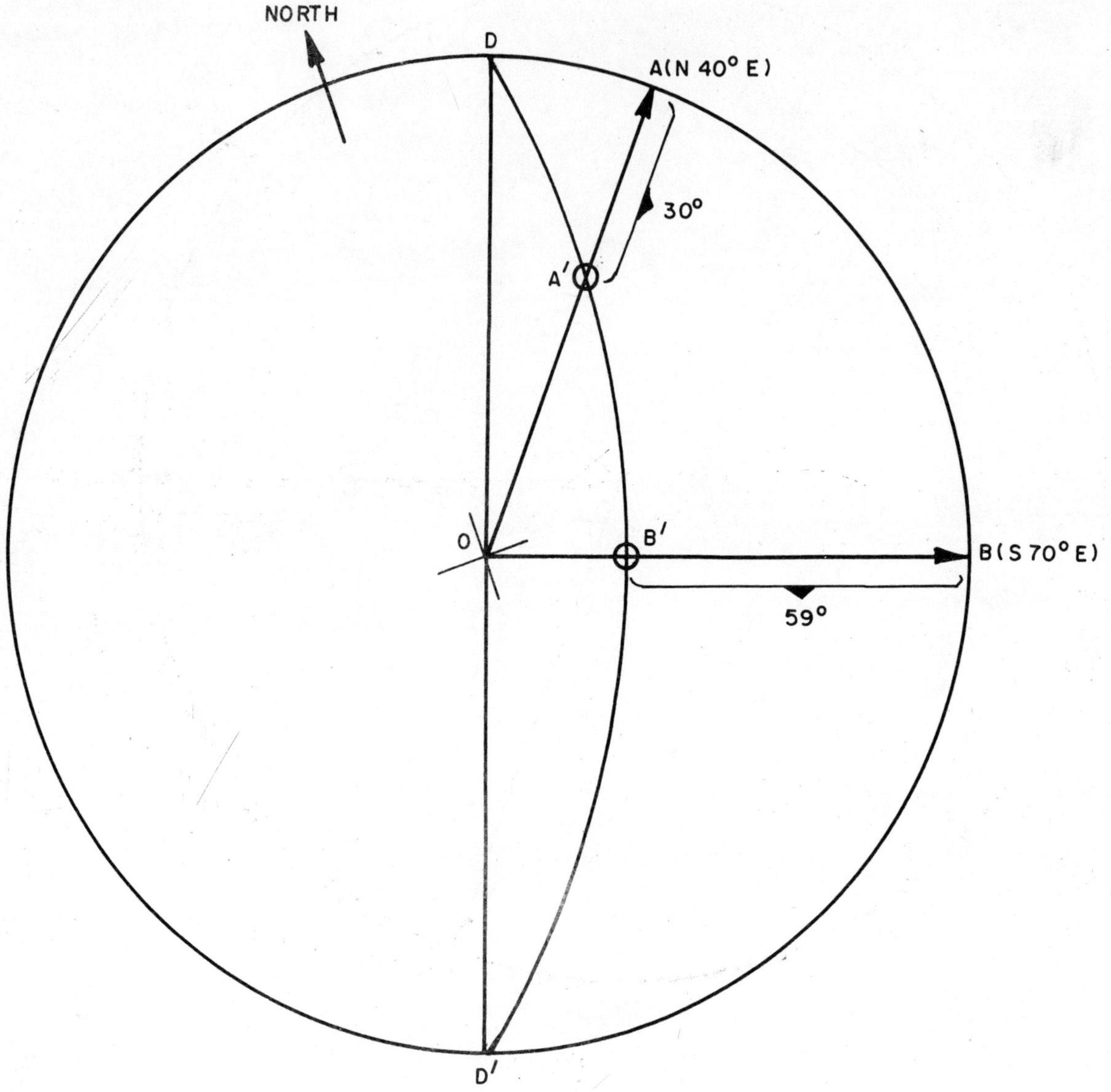

Fig. 5.3.3

Fig. 5.3.3 In this position arc of great circle passing through point *A′* is drawn, which cuts *OB* at point *B′*. *BB′* is found to be 59°. This is the amount of true dip along *OB′B* (S 70 °E)

tracing sheet to zero position which gives the direction of true dip as S 40 °E (line *OB′*B Fig. 5.3.7). It is not necessary to place the line *OA′A* only in the eastern side of the stereographic net as shown in Fig. 5.3.6. The position of *OA′A* may be taken on the western side of the net. The ultimate result of true dip amount or the direction of true dip, is not affected. The amount of true dip is found to be 53° and the direction of dip to be S 40 °E.

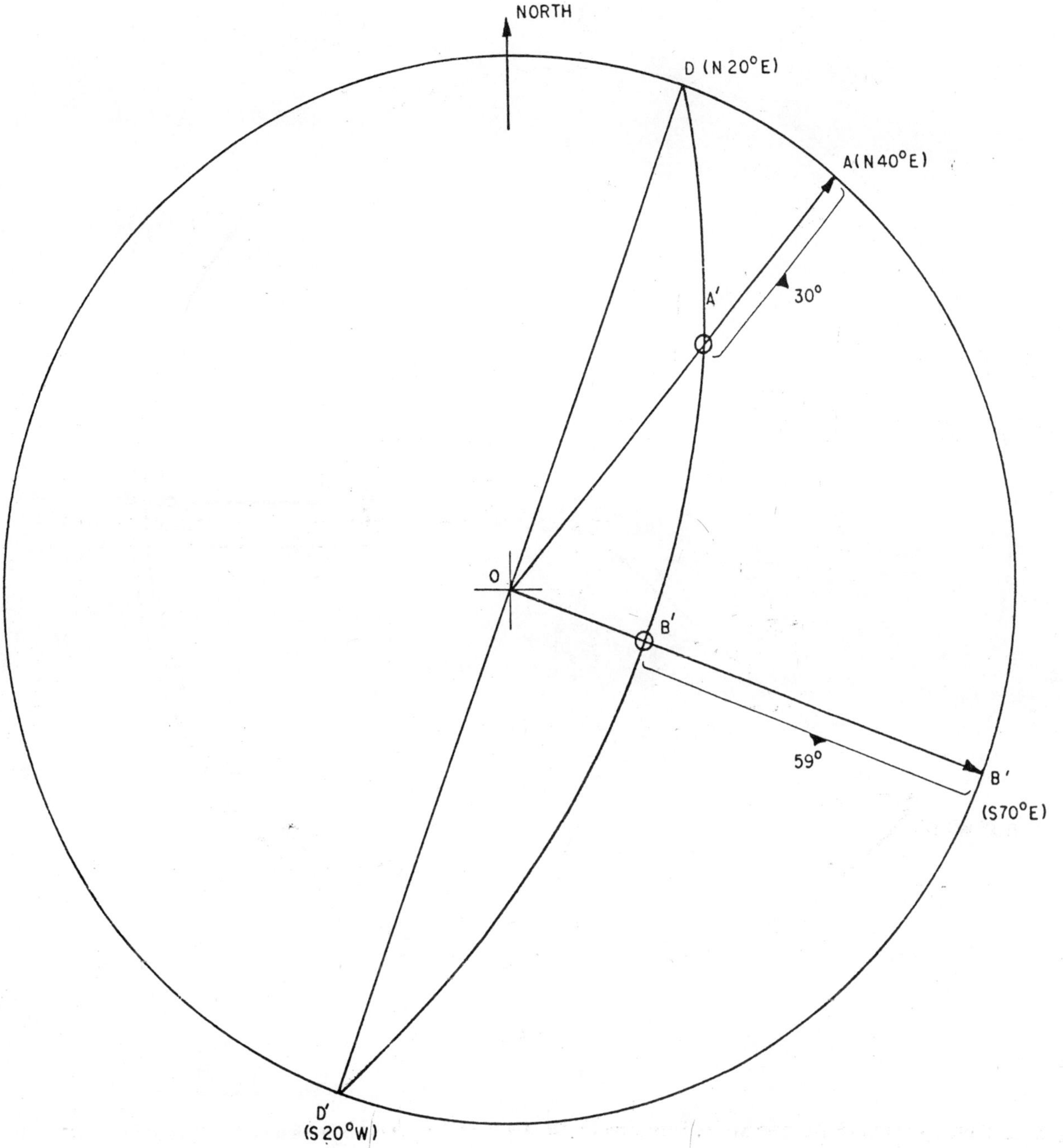

Fig. 5.3.4

Fig. 5.3.4 Derivation of True Dip Amount, Given its Direction, and one Apparent Dip (Problem 5.5)

NOTE : This is the zero position of the stereogram. *DOD′* gives the direction of strike (N 40°E - S 40 °W). The amount of true dip (59°) is read off in Fig. 5.3.3.

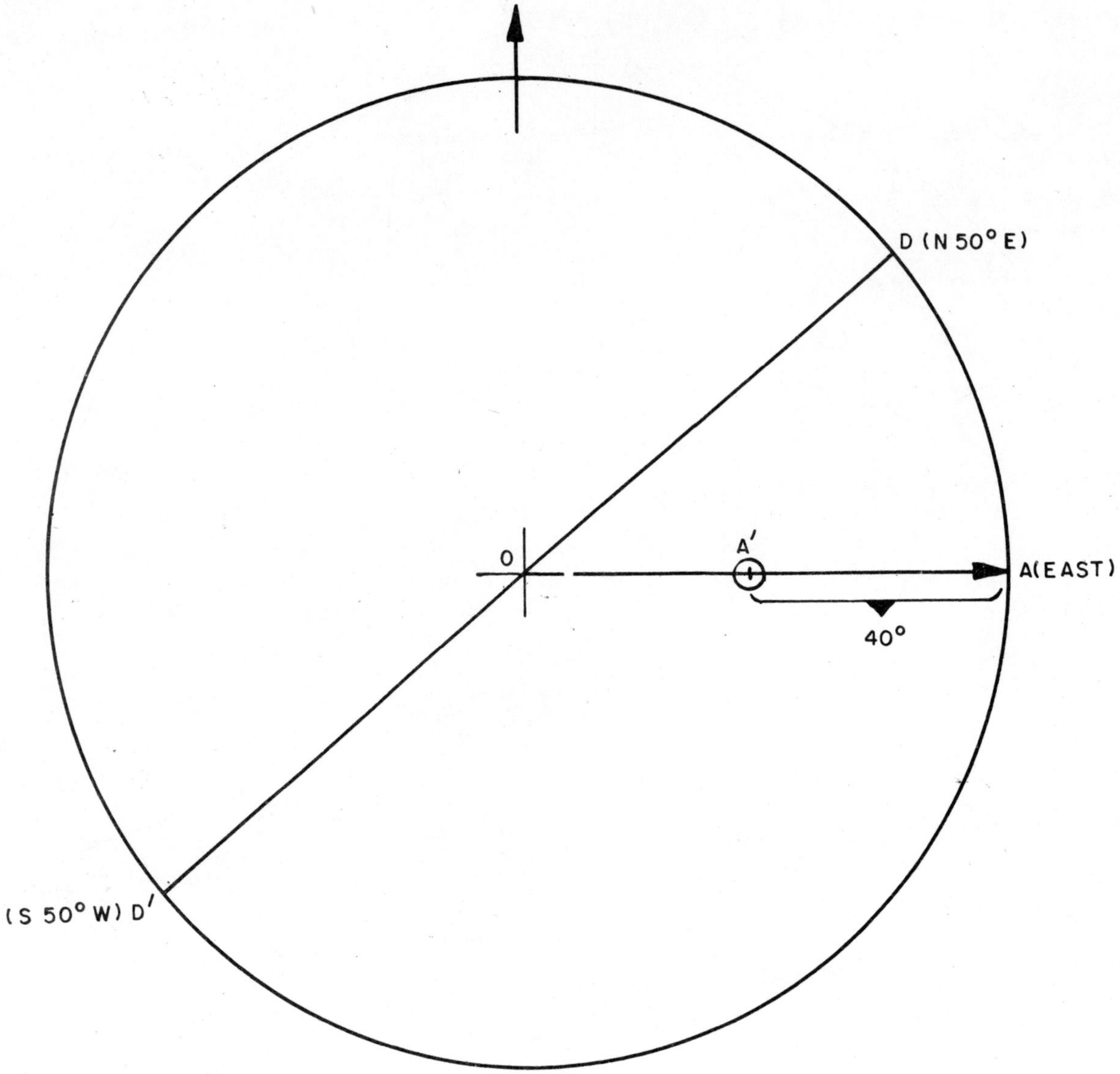

Fig. 5.3.5

Fig. 5.3.5 **Plotting of OA (East) and Strike Direction (N 50 °E - S 50 °W) (Problem 5.6)**

NOTE : This is done in zero position of the stereogram. *DOD′* is the direction of true strike (N 50 °E – S 50 °W) and *OA* is the direction of apparent dip (East). The amount of apparent dip of 40° also is plotted and point *A′* is obtained.

(d) One Apparent Dip and Amount of True Dip

Problem 5.7 A formation has an apparent dip of 30° due S 60 °W. If the true dip amount be 50°, find the direction of true dip.

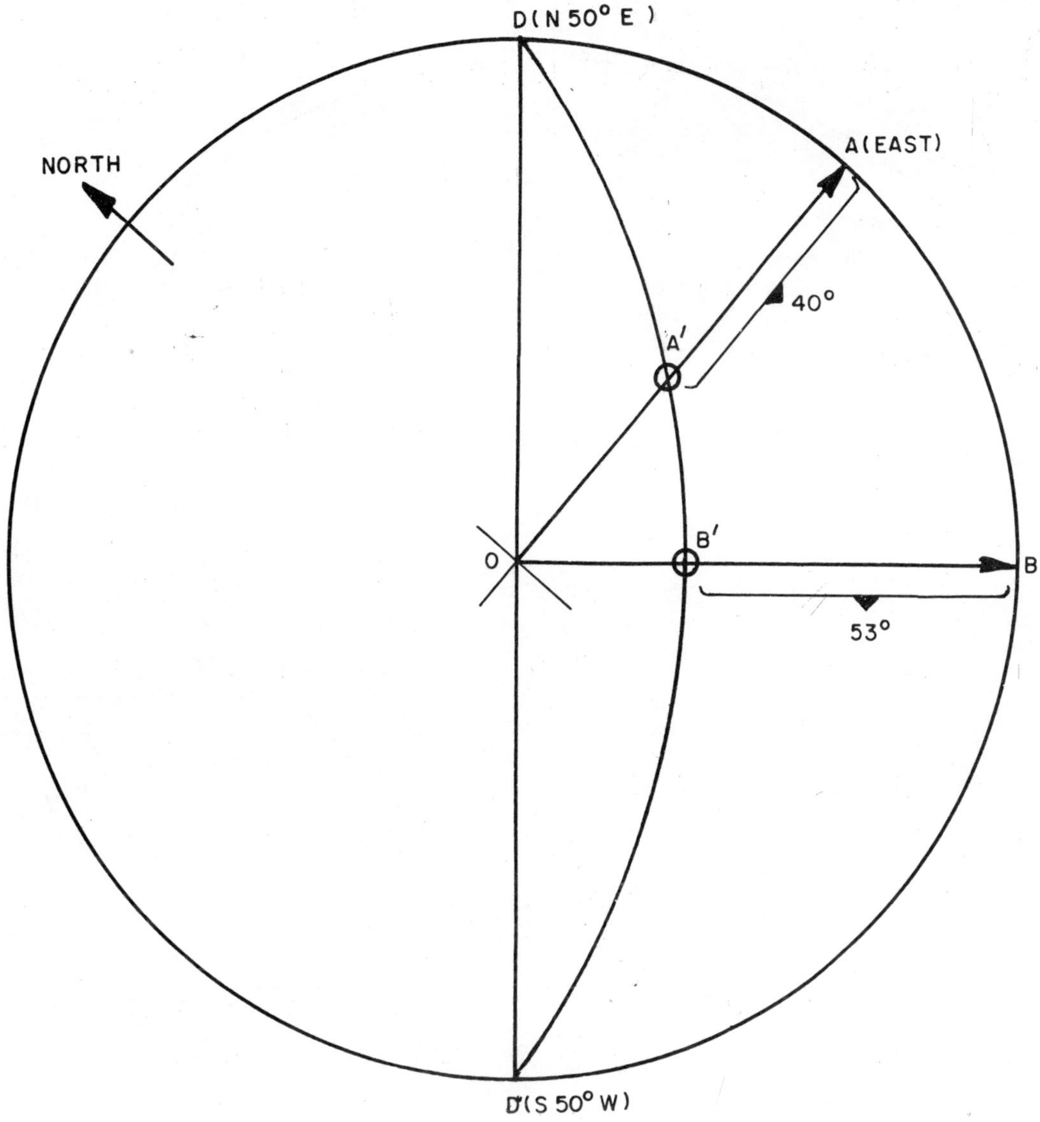

Fig. 5.3.6

Fig. 5.3.6 Rotating *DOD'* to Coincide with N-S Diameter of the Net (Problem 5.6)

NOTE : In this position, the arc of great circle passing through point *A'* is drawn. The eastern radius also is drawn and is labelled as *OB*, which cuts the arc of great circle at *B'*. Measure *BB'* which gives the amount of true dip along *OB'B*, but its direction is to be read off only after rotating the tracing paper to zero position.

Procedure

Draw *OA* due S 60°W and mark off 30° on it to obtain point *A'* (Fig. 5.3.8 and 5.3.9). Rotate the tracing sheet till the point *A'* is seen to fall on arc of great circle of 50° value (Fig. 5.4.0). Trace the said arc of great circle upto points *D* and *D'*. Trace

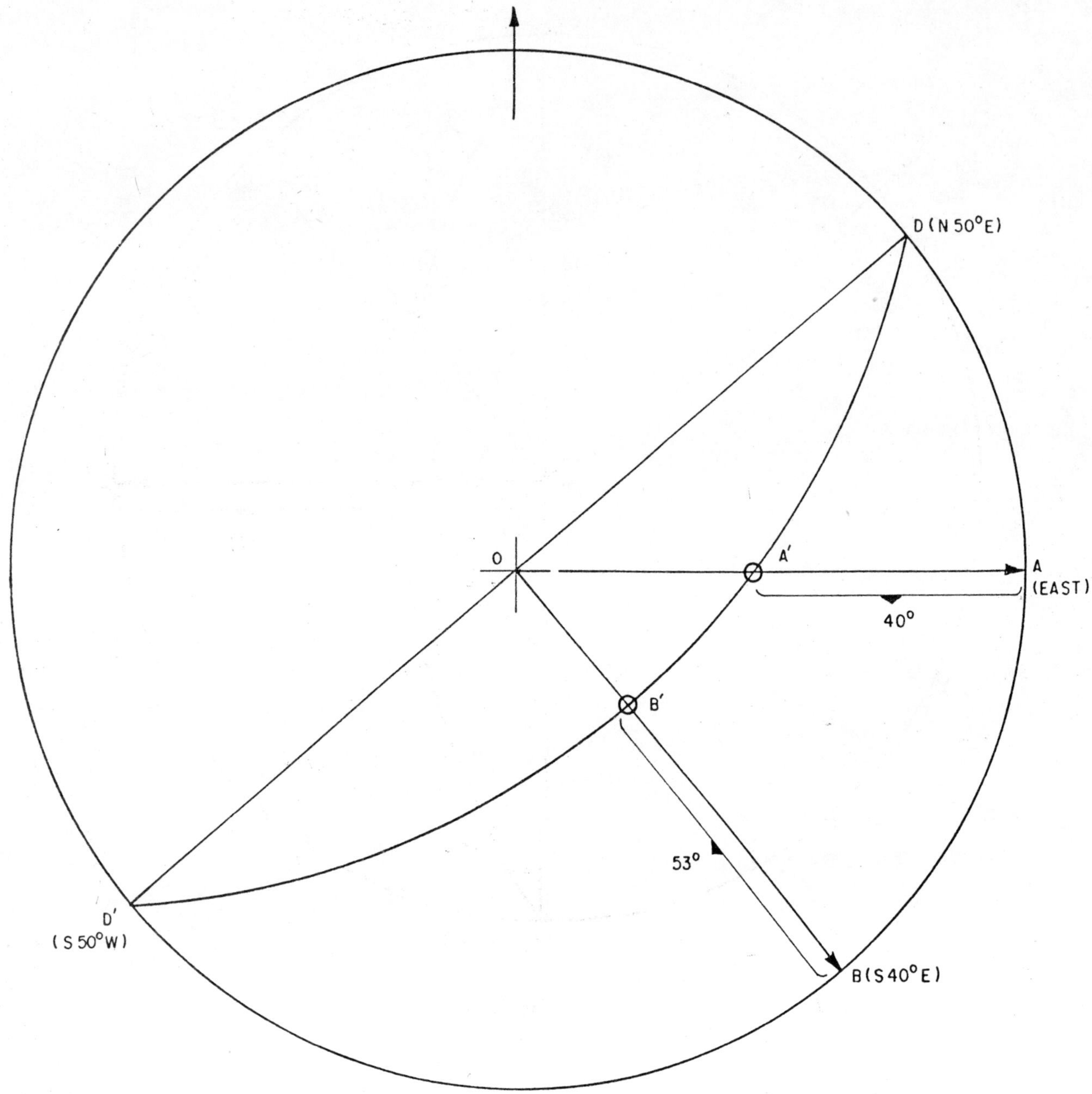

Fig. 5.3.7

Fig. 5.3.7 Derivation of Direction and Amount of True Dip, from one Apparent Dip and True Strike Direction (Problem 5.6)

NOTE : Read off the position. taken by *OB′B* when the tracing paper is brought to zero position. It is found to be S 40 °E. The amount of true dip along *OB′B* is read off in Fig. 5.3.6

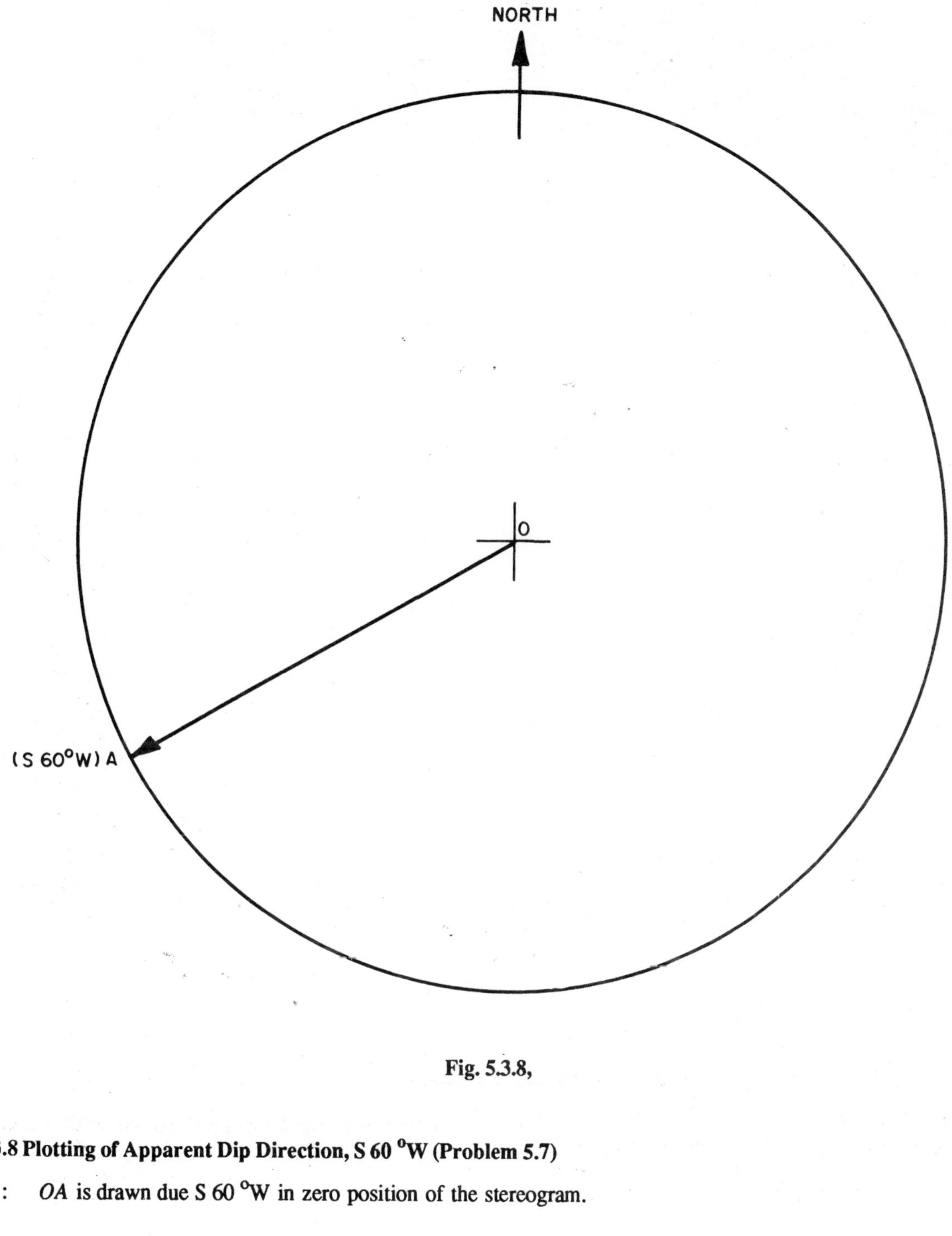

Fig. 5.3.8,

Fig. 5.3.8 Plotting of Apparent Dip Direction, S 60 °W (Problem 5.7)

NOTE : *OA* is drawn due S 60 °W in zero position of the stereogram.

the radius which gives the direction of true dip (line *OB* in Fig. 5.4.0) and which is one of the directions of true dip. The line *OB* intersects the arc of great circle at point *B'*. There will be yct another poisition of point *A'* when it again falls on arc of great circle having a value of 50°. Therefore rotate the tracing sheet again and obtain the second position as shown in Fig. 5.4.1. Trace the arc of great circle upto point *E* and *E'*, and trace the radius *OC* (Fig. 5.4.1). In so doing point *C'* is obtained. Line *OC'C* is the second direction of true dip. Rotate the tracing sheet to zero position and read off directions

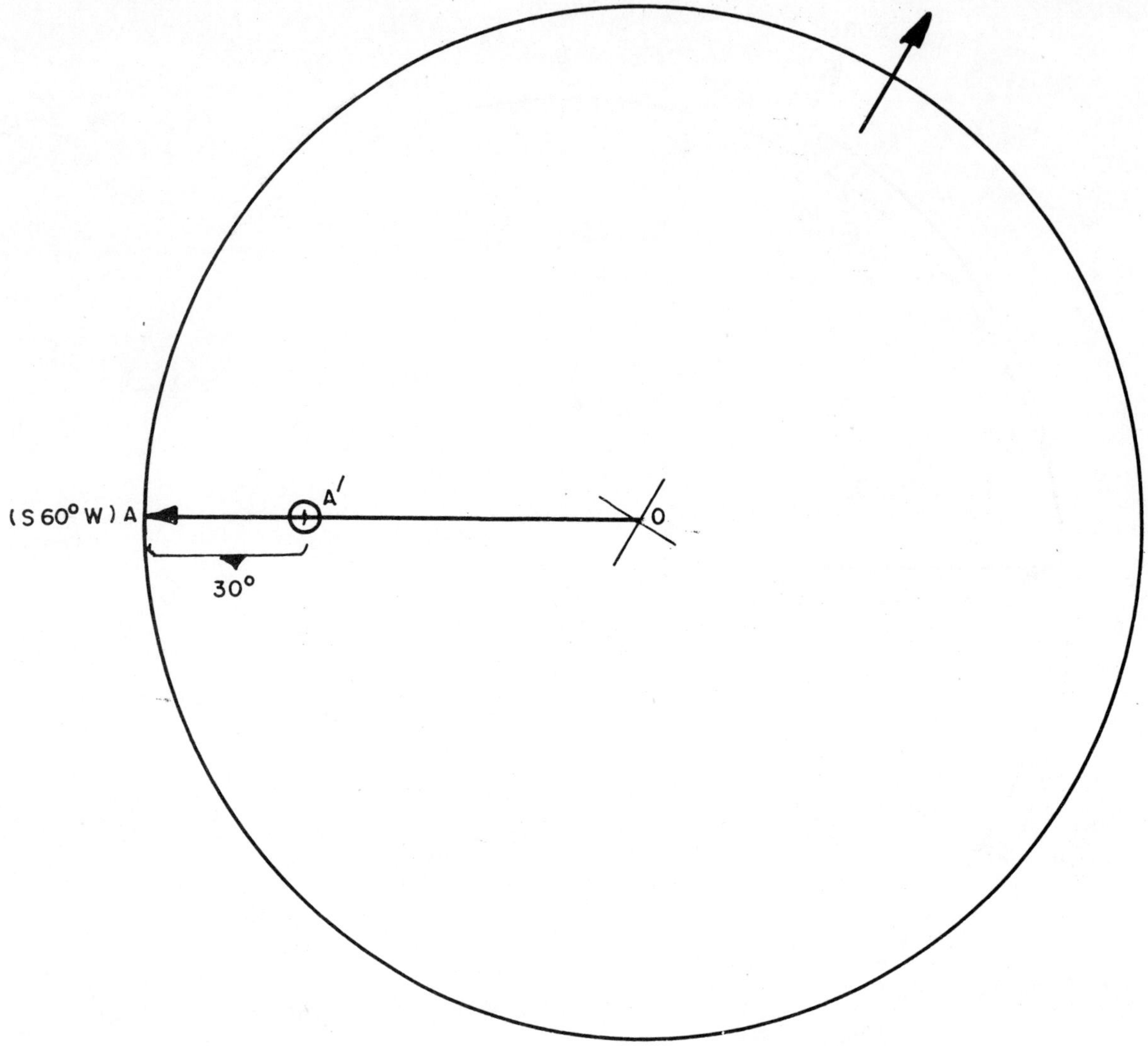

Fig. 5.3.9

Fig. 5.3.9 Rotating *OA* to Coincide with Western Radius (Problem 5.7)

NOTE : *OA* is made to coincide with western radius and point *A'* obtained on it such that *AA'* is of value 30° which is the amount of apparent dip along *OA* (S 60 °W).

OB', B, OC', C, DOD', and *EOE',* which give the directions of true dip and of true strike. (Fig. 5.4.2). The directions of true dip are found to be N 60 °W and south, the corresponding strike directions being N 30 °E–S 30 °W and E–W, respectively.

5.1.9 Bearing and Plunge of Line of Intersection of two Planes

Introduction. When surfaces intersect each other, it gives rise to a line of contact. Such a line has therefore its trend and inclination. The inclination of a line is called as *"plunge"*. The trend of the line of intersection is called *"bearing"*. Thus

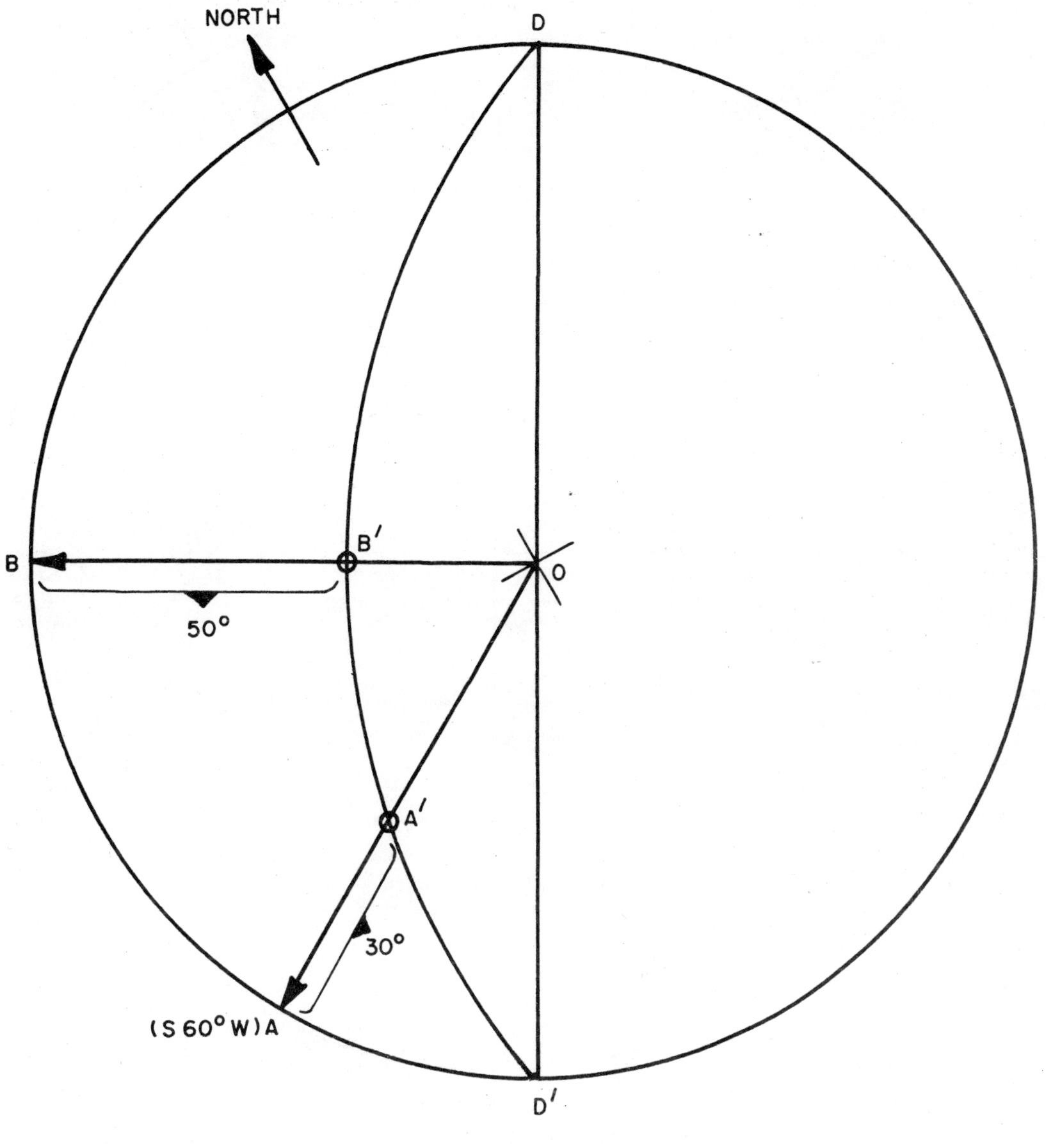

Fig. 5.4.0

Fig. 5.4.0 Rotating *OAA* and Positioning *A* on 50° Arc of Great Circle (Problem 5.7)

NOTE : Tracing paper is rotated till *A′* is found to fall on 50° arc of great circle. The arc is drawn, the western radius is also drawn and is labelled as *OB*. The arc cuts the line *OB* at *B′*. *BB′* is found to be of value 50° and *OB′B* is one of the true dip directions, its direction is to be read off in the zero position of the stereogram. Draw the diameter *DOD′* which is the true strike direction.

the two limbs of a fold meet in a line, which is designated as the axis of the fold. The axis therefore has trend and an inclination. Many times the intersection of geological bodies, marks a zone of active mineralization of economic minerals. The attitude of such a zone is definitely required to be established for the purpose of its later mining. Keeping

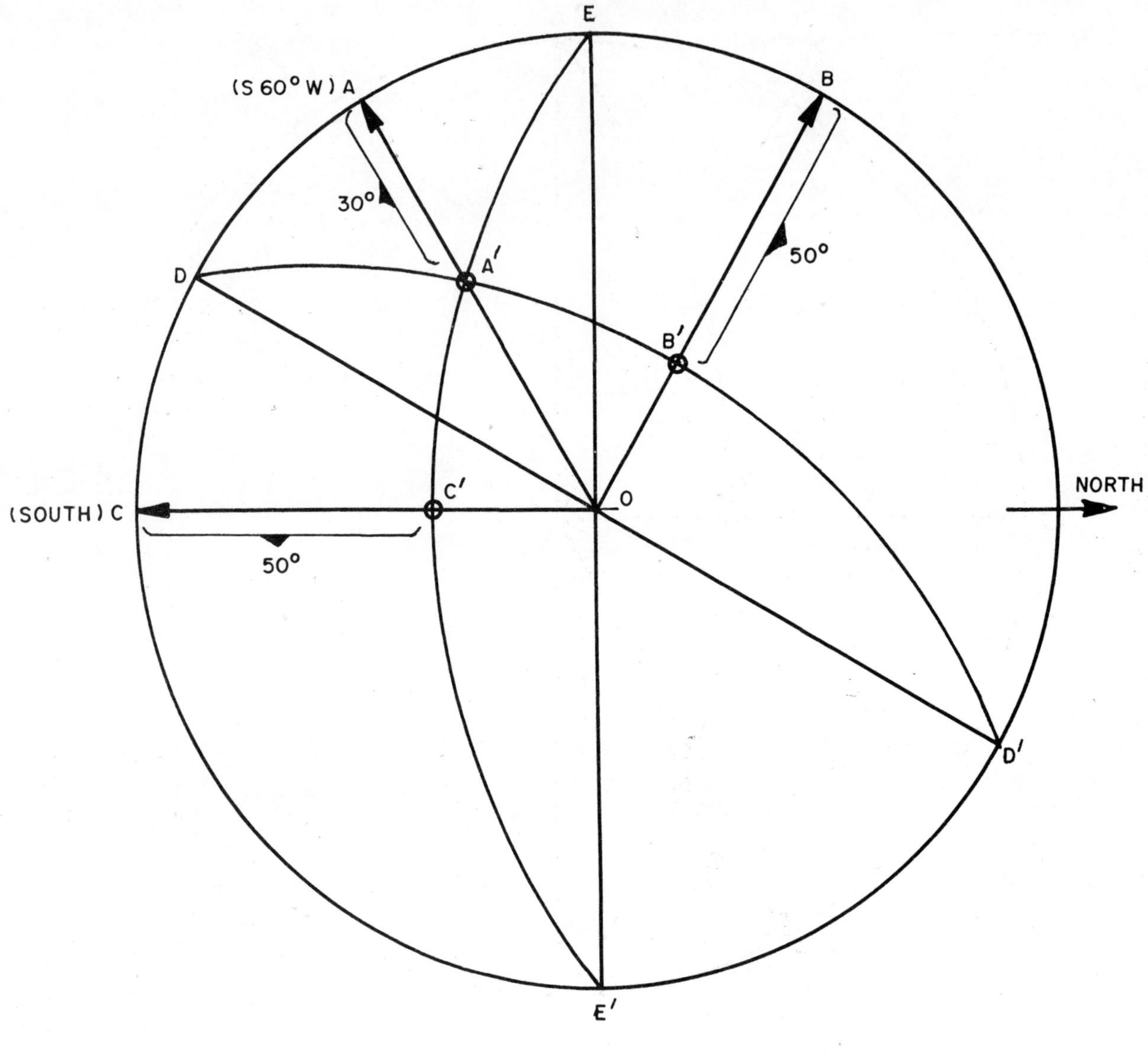

Fig. 5.4.1

Fig. 5.4.1 **Rotating $OA'A$ again and Bringing A' to Fall again on Arc of 50° (Problem 5.7)**

NOTE : Point A' can be made to fall on arc of great circle of 50° again which is different from that obtained in Fig. 5.4.0. This new arc of great circle is drawn through point A' and the western radius is drawn and labelled as OC. The line OC cuts the arc of great circle at C' and CC' will be found to be of 50° value. $OC'C$ is yet another true dip direction, but its direction is to be read off only in the zero position of the stereogram. Draw the diameter EOE', which is the true strike direction.

these features in view, in the following paragraphs the determinations of the bearing and the plunge of the line of intersection of two dipping surfaces is described.

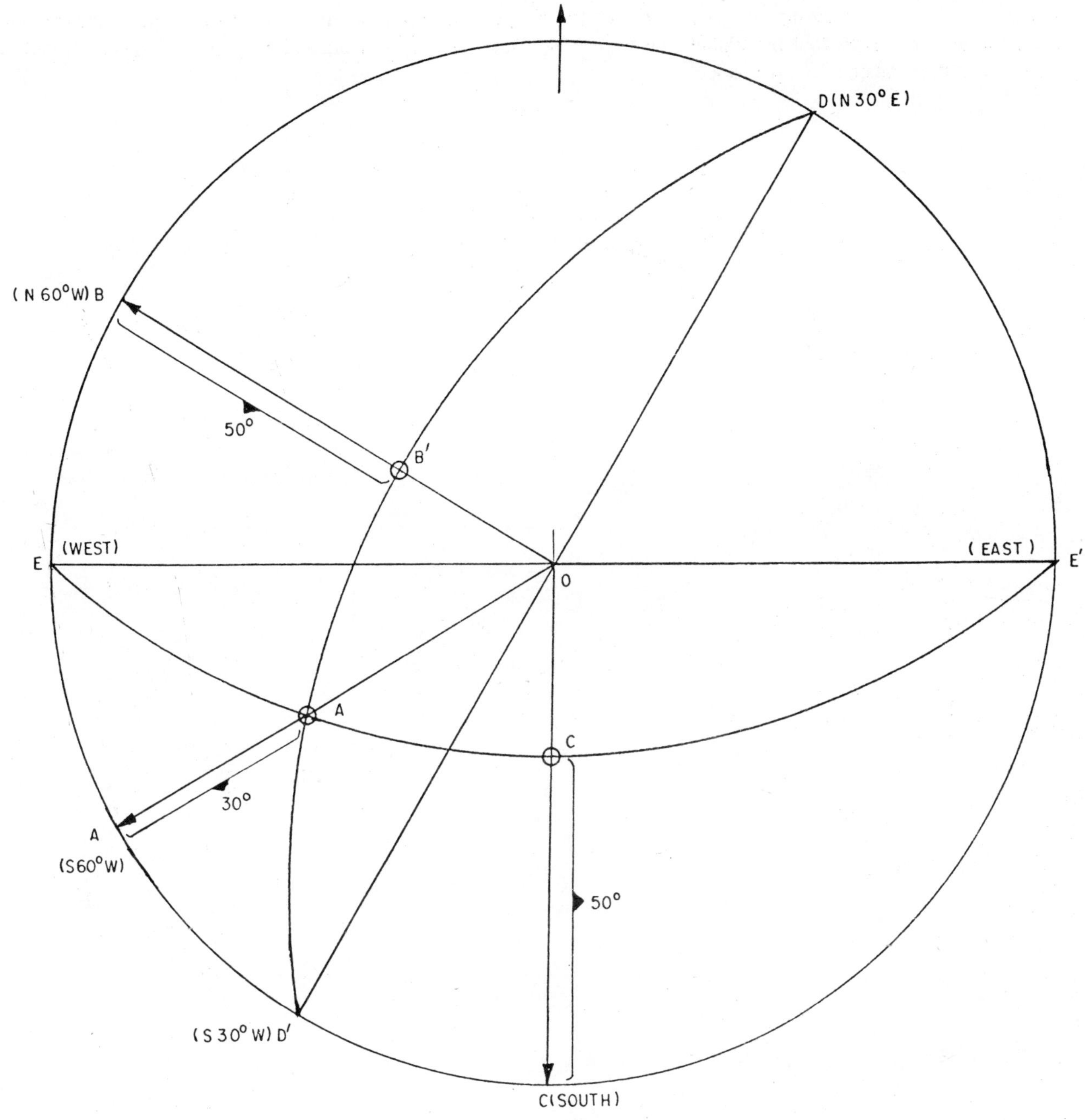

Fig. 5.4.2

Fig. 5.4.2 Derivation of True Dip Directions from one Apparent Dip and True Dip Amount (Problem 5.7)

NOTE : The stereogram is brought to zero position and the directions of true dip *(OB′B* and *OC′C)* and the direction of true strikes *(DOD′* and *EOE′)* are read off. *OB′B* and *OC′C* are found to be N 60 °W and south, respectively. *DOD′* and *EOE′* are found to be N 30 °E - S 30 °W and E-W, respectively. Note also that if only true dip amount is given, then there are two directions of true dip.

Problem 5.8 A sandstone dips 40° due N 40 °E and is cut by a fault dipping 50° due N 60 °W. The line of intersection of these two planes marks a zone of sulphide mineralization. Determine the bearing and the plunge of the line (zone) of intersection of the sandstone and the fault.

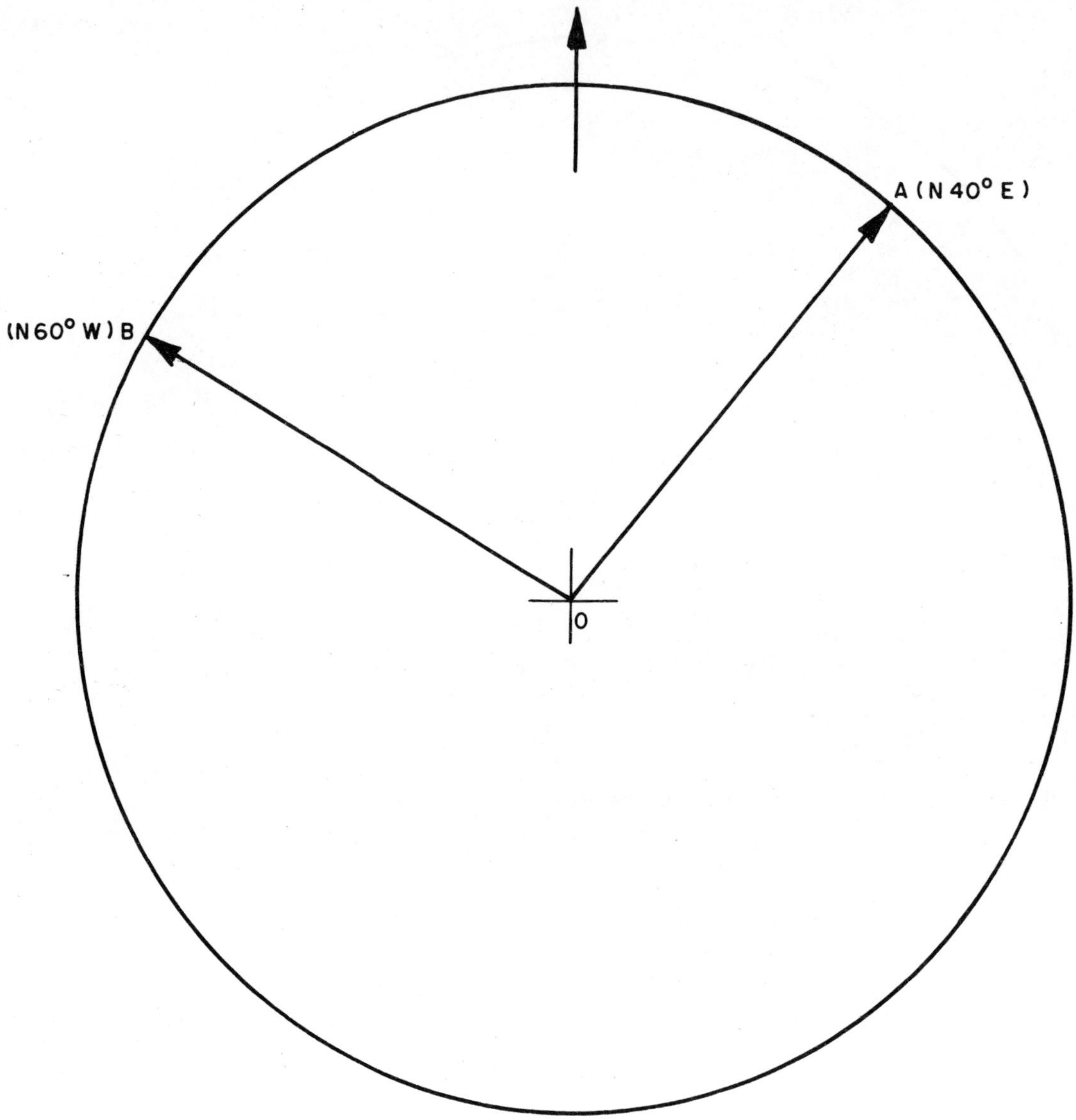

Fig. 5.4.3,

Fig. 5.4.3 **Plotting Directions N 40 °E - (*OA*) and N 60 °W (*OB*** (Problem 5.8)

NOTE : *OA* and *OB* are drawn in zero position of the stereogram.

Procedure

Draw lines OA and OB due N 40 °E and N 40 °W, respectively (Fig. 5.4.3). Bring OA to coincide with the eastern radius and obtain point A' such that AA' is equal to 40°. Now draw the arc of great circle through point A' as shown in Fig. 5.4.4.

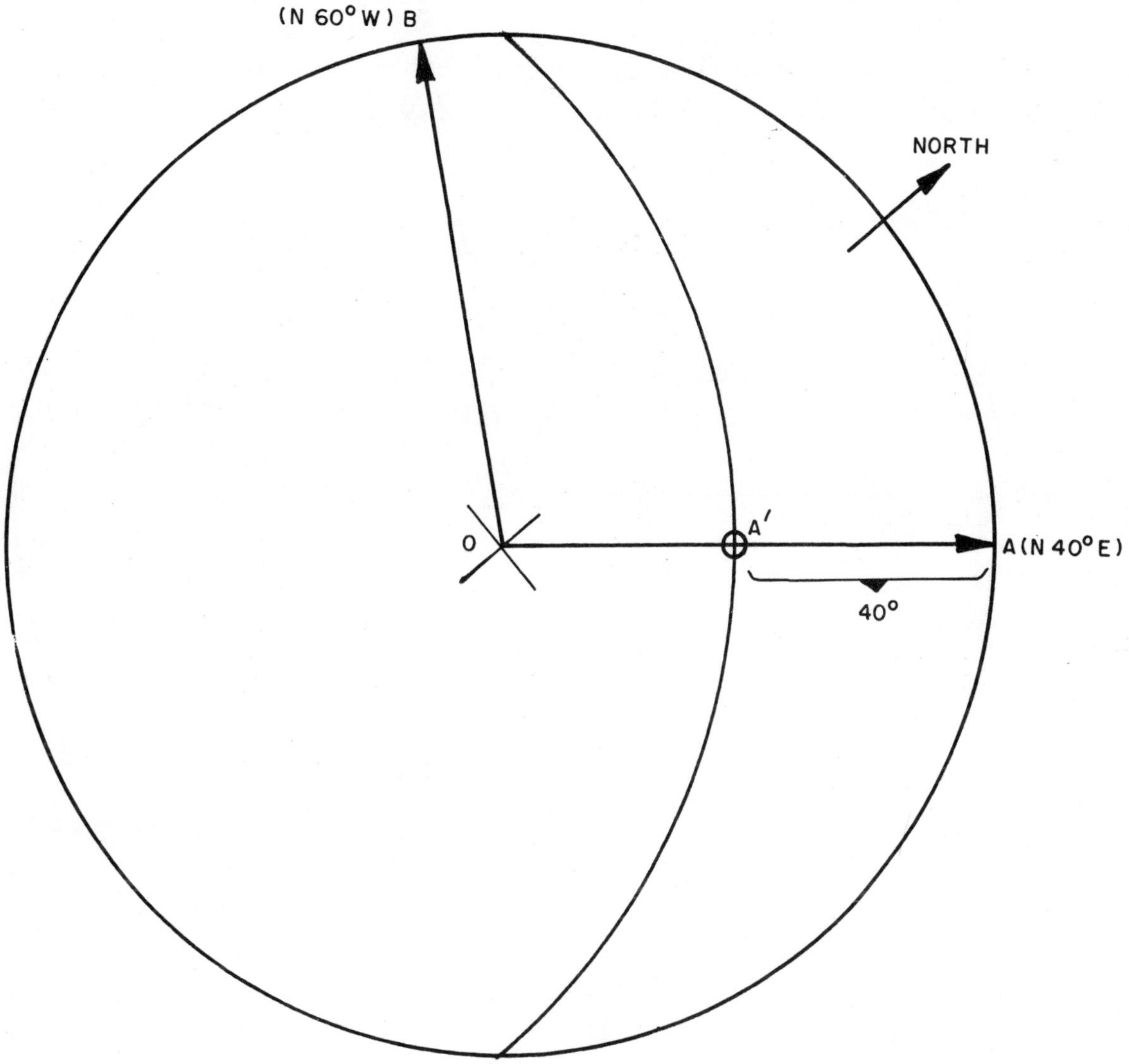

Fig. 5.4.4

Fig. 5.4.4 Rotating OA to Coincide with Eastern Radius (Problem 5.8)

NOTE : OA is made to coincide with the eastern radius and point A' obtained on it such that AA' is of 40° value. Draw arc of great circle through A'. This arc of great circle stands for fault plane.

Similarly bring the line *OB* to coincide with the western radius and mark off an angle of 60° and obtain point B'. Draw the arc of great circle through point *B'* as shown in Fig. 5.4.5. The two arcs of great circle intersect at point *C'* (Fig. 5.4.5). Join

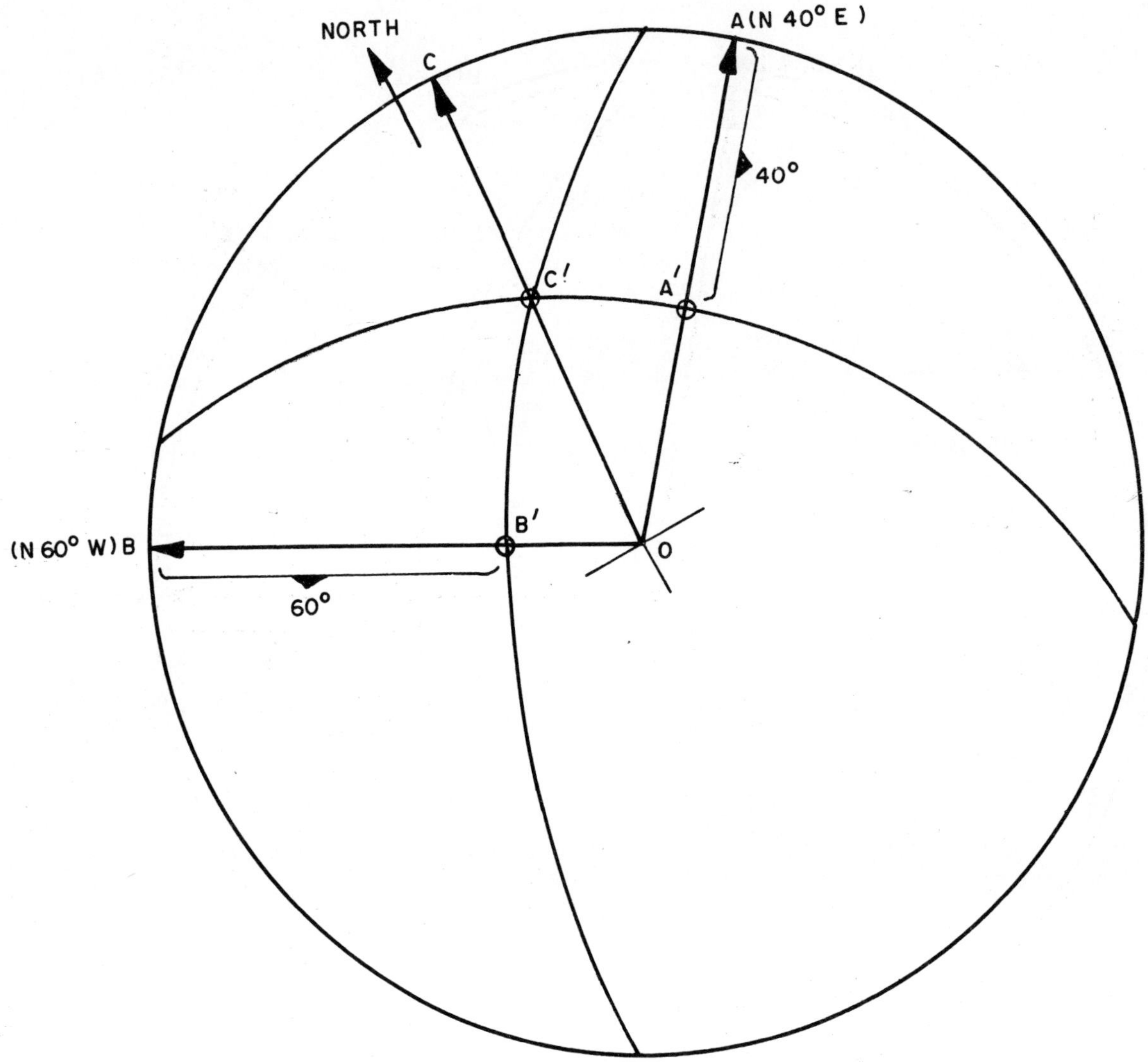

Fig. 5.4.5

Fig. 5.4.5 Rotating *OB* to Coincide with Western Radius (Problem 5.8)

NOTE : *OB* is made to coincide with western radius and a point *B'* is obtained on it such that *BB'* is of 50° value. Draw arc of great circle through *B'*. This arc of great circle stands for dipping surface of sandstone. Arcs of great circles drawn through *A'* and *B'* intersect at *C'*. Join *OC'* and produce it to *C*. *OC'C* is the line of intersection of fault and standstone. Angular distance *CC'* stands for the plunge of the line *OC'C*.

OC' and extend it upto C. The line $OC'C$ is the line of intersection of sandstone and fault plane. Rotate the tracing sheet and bring $OC'C$ to coincide with the eastern radius and measure angle CC' (Fig. 5.4.6). CC' is found to be 35^0 which is the plunge of the line of intersection. Now bring the tracing sheet to zero position and read off the bearing of line $OC'C$, which is found to be N 7 °E (Fig. 5.4.7).

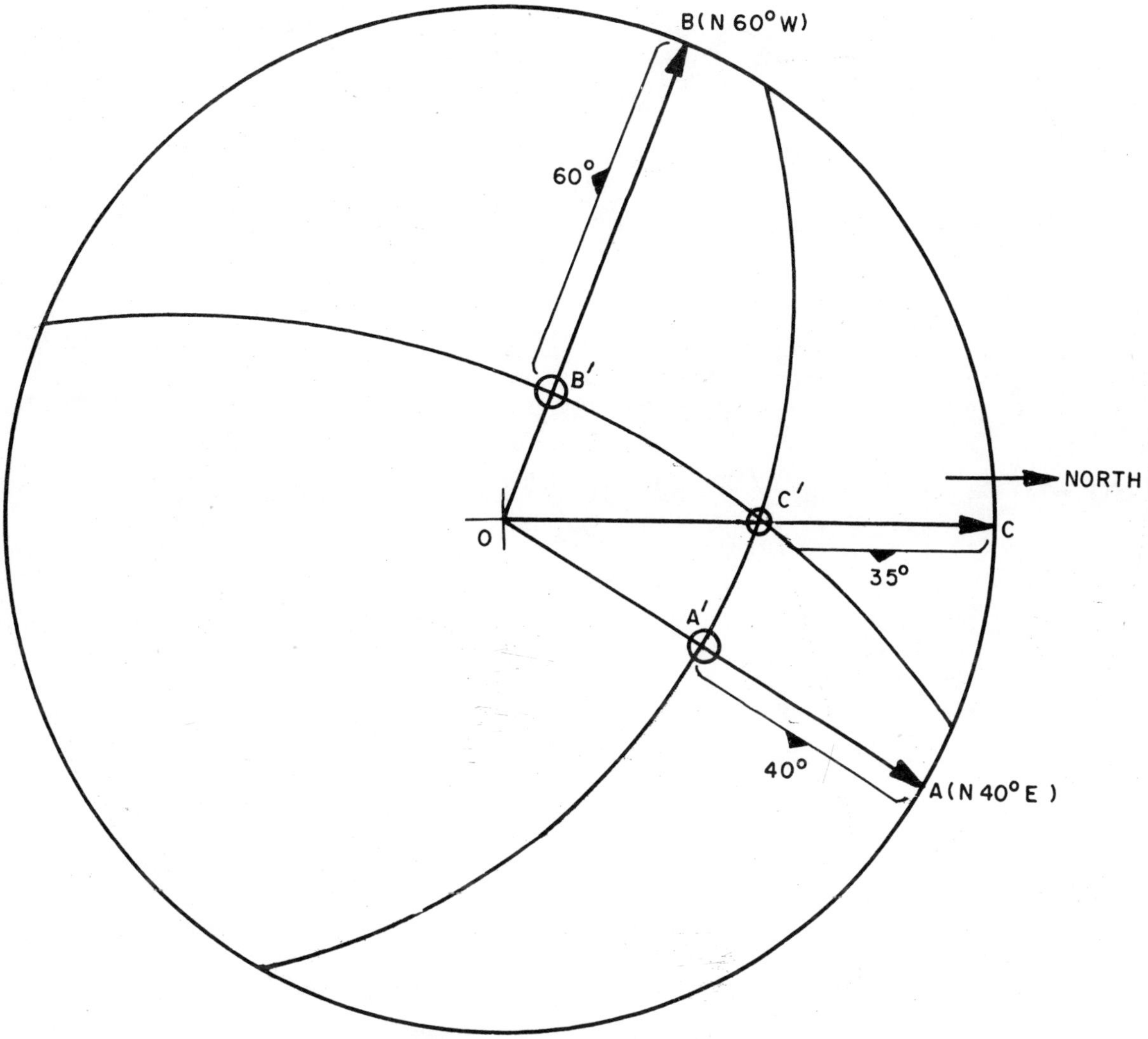

Fig. 5.4.6

Fig. 5.4.6 Rotating $OC'C$ to Coincide with Eastern Radius (Problem 5.8)

NOTE : The tracing paper is rotated and $OC'C$ is made to coincide with eastern radius. The distance CC' is read off which is found to be of 35° value. This is the plunge of the line of intersection of fault and dipping surface of sandstone.

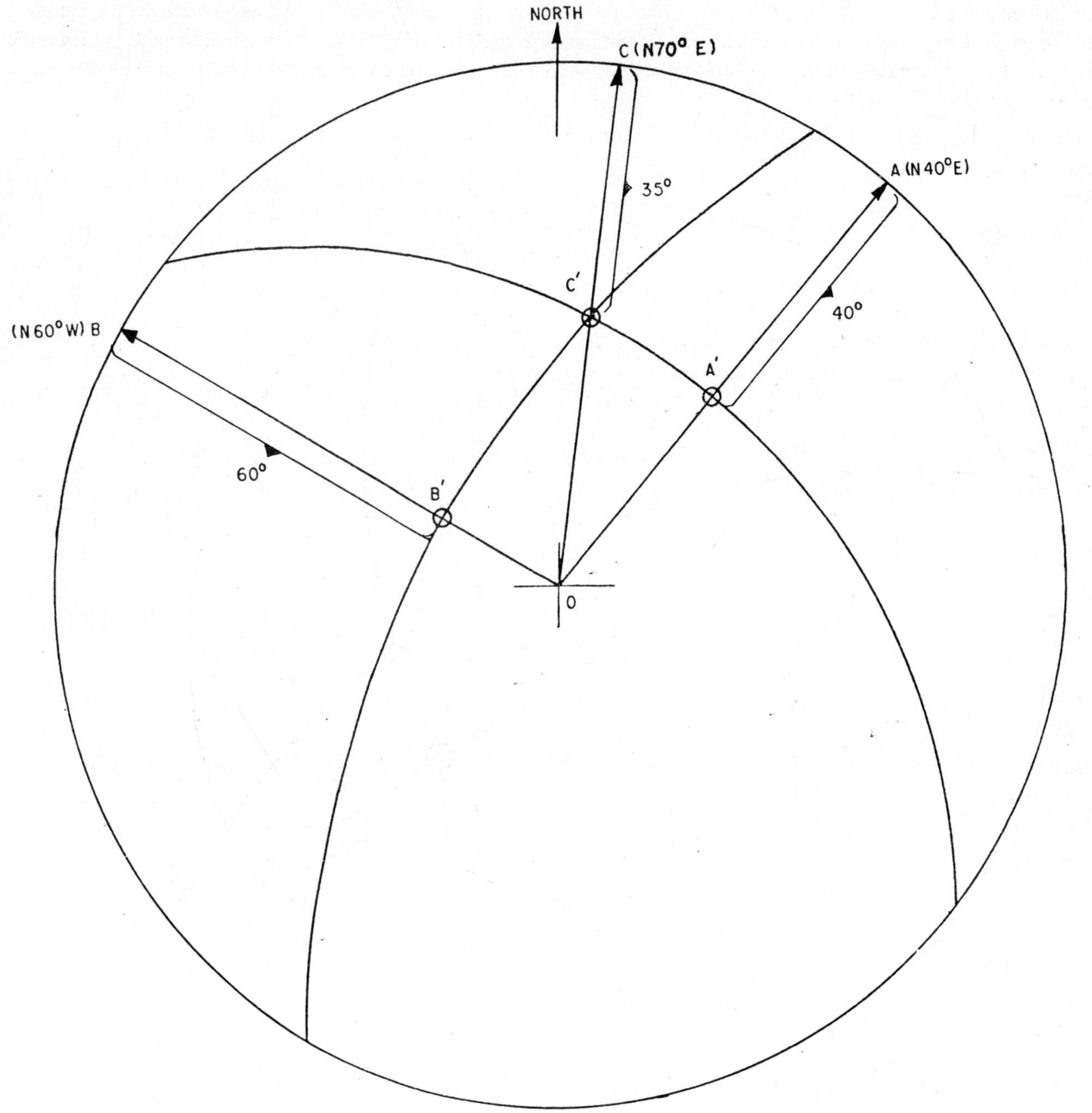

Fig. 5.4.7

Fig. 5.4.7 Derivation of Bearing and Plunge of Line of Intersection of Fault Plane and Sandstone (Problem 5.8)

NOTE : Stereogram is brought to zero position and direction of *OC'C* is read off. It is found to be N 7 °E, which is its bearing. The plunge of this line *OC'C* is read off in Fig. 5.4.5, which is found to be of 35° value.

6

USE OF 'H' STRUCTURE CONTOURS IN GEOLOGICAL PROBLEMS

6.1.1. Introduction

The bearing and the plunge of the line of intersection of two dipping planes can also be determined by the geometrical construction. The data may be set in the form of a sketch or it may be stated in words. If the data be stated in text, then a sketch is required to be drawn. If it be already in the form of a sketch, the further constructional steps are immediately adopted. This method is especially extendable to determine the downthrow side of a fault, the trend of net slip and also the rake of net slip. The methodology is described below.

6.1.2 Determination of Bearing and Plunge of Line of Intersection

(a) Data Given in Degrees

Problem 6.1. A sandstone dips 20° due N 40 °E and is cut by a fault dipping 50° due N 60 °W. Determine the bearing and the plunge of the line of intersection of these two planes by the geometric method.

Procedure

Step I Drawing Sketch of Intersection of the two Surfaces

As a first step, the trend (meaning the direction of strike) of the two dipping surfaces (sandstone and fault in the present case) are drawn and are made to intersect each other. As the directions of dip are given, the trends will be perpendicular to these i.e., for the sandstone it will be N 50 °W– S 50 °E (the dip is due N 40 °E), and for the fault plane it will be N 30 °E– S 30 °W (the dip is due N 60 °W). These trends are shown by the lines AOA' and BOB' for the sandstone and fault plane, respectively. (Fig. 6.1.0). The direction of dips are also shown by symbols. The intersecting lines AOA' and BOB' produce four sectors, out of which that sector towards which both the surfaces are dipping is required to be established (Sector I in Fig. 6.1.0). This is necessary because from the first intersection of the two plane surfaces (point O in Fig. 6.1.0), the direction in which the second point of intersection will be encountered at depth, can be reckoned. From plotting of the trends of the sandstone and the fault which also indicates the direction of dip symbolically, it is possible to ascertain that the second point of intersection of the dipping surfaces will be generally due north (Sector I Fig. 6.1.0). This inference is drawn from the disposition of the dip direction of both sandstone and the fault being towards Sector I.

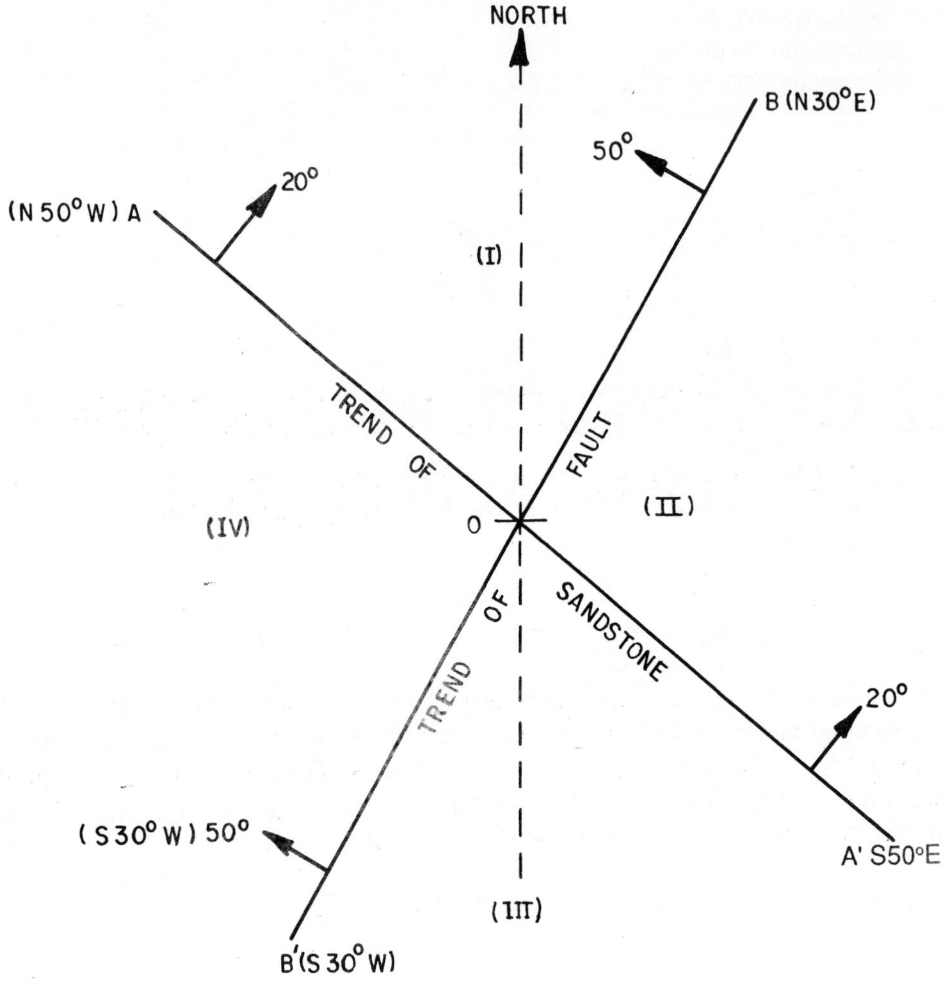

Fig. 6.1.0

Fig. 6.1.0 PLOTTING OF TRENDS, DIRECTIONS OF DIP AND AMOUNT, OF THE SANDSTONE AND FAULT (Problem 6.1)

NOTE In such plotting the trends meaning the directions of strike of the dipping body are needed. If the dip direction alone be given, then the direction of strike should be derived and plotted. The direction of dip and amount are also shown in all the four sectors. Note that the two surfaces dip towards the first sector, which gives the region in which the second point of intersection of the surface is to be expected. The later construction of "H" structures are to be effected in sector I.

Step II Construction of "*H*" Structure Contous

The two lines drawn in Fig. 6.1.0 (*AOA'* and *BOB'*) stand for the outcrop of the two surfaces. Naturally these outcrops will get obscured in the dip direction, because these will be taken to depth with increased horizontal distance and hence these cannot crop out again at the surface. However, from the knowledge of dip amount and the direction, the level reached by the two dipping surfaces could be calculated. From such a knowledge, the second point of intersection (the first is at *O* in Fig. 6.1.0) can be ascertained. But this second point will be at a level lower than the first point *O*. Now since the two points on the line of intersection of the two dipping surfaces are available, it is possible to calculate the bearing and plunge of the said line of intersection. The second point of intersection is obtained by constructing "*H*" structure contours (Fig. 6.1.1).

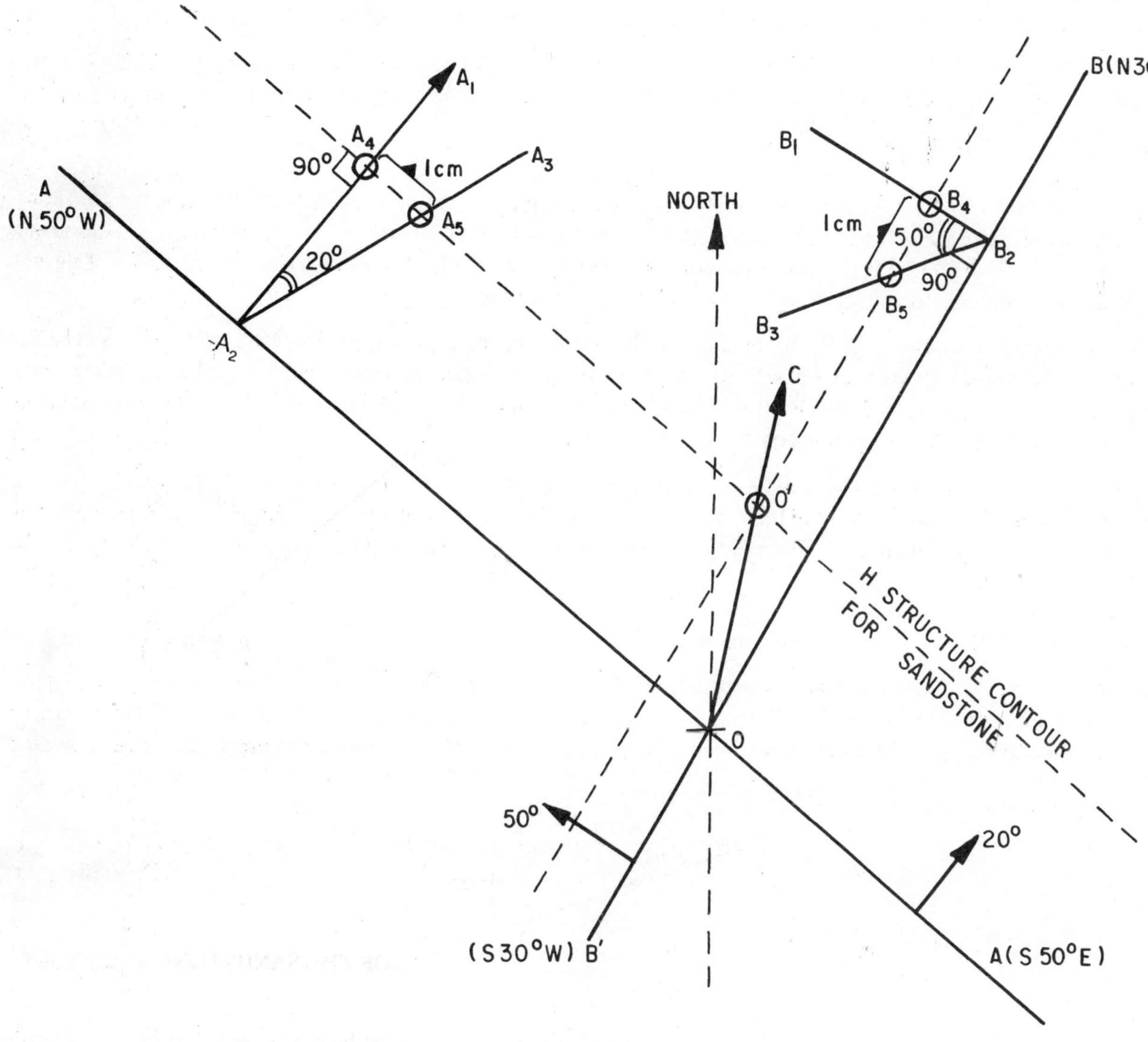

Fig. 6.1.1

Fig. 6.1.1 Construction of "H" Structure Contours for Sandstone and Fault Planes (Problem 6.1)

NOTE : *AOA'* - trend of sandstone = N 50 °W – S 50 °E

A_2A_1 = perpendicular on AOA' at A_2

$A_1A_2A_3$ = 20°, A_4–A_5 = perpendicular on A_1A_2 at A_4 from A_2A_3 ,length of A_4–A_5 = "*H*" depth = 1 cm

The dashed line drawn through points A_4 and A_5 and parallel to AOA' is the "H" structure contour of sandstone at depth of 1 cm (*H* depth). This is the projection of surface of sandstone occurring at "*H*" depth, on horizontal surface.

BOB' = trend of fault plane = N 30 °E - S 30 °W

B_2B_1 = perpendicular on BOB' at B_2

$B_1B_2B_3$ = 50°, B_4B_5 = perpendicular on B_1B_2 at B_4 from B_2B_3, length of B_4B_5 = "*H*" depth = 1 cm

The dashed line drawn through points B_4 and B_5 and parallel to BOB' is the "*H*" structure contour of fault plane at depth of 1 cm (*H* depth). This is the projection of fault plane occurring at "*H*" depth, on horizontal surface.

The dashed lines intersect at O'.O' is the second point of intersection of sandstone and fault at depth "*H*". $OC'C$ is the line of intersection of the two surfaces. Its bearing is found to be N 12 °E. The plunge of this line is measured separately.

Erect a perpendicular A_1–A_2 at A_2 on the line AOA'. Construct an angle A_1 A_2 A_3 of a value 20°, because the sandstone has this dip amount. Erect a perpendicular A_4–A_5 on A_1A_2 at A_4 such that the length of A_4– A_5 will be some convenient value (1 or 2 cms). Extend this line (A_4–A_5) on either sides which will be found to run parallel to line AOA'. This line A_4–A_5 is the "*H*" structure contour (shown as a dashed line) of the sandsone (Fig. 6.1.1.).

Similarly erect a perpendicular B_1–B_2 on BOB' at B_2. Construct an angle B_1 B_2 B_3 of value 50°, because the dip of fault plane is of that value. Erect a perpendicular B_4–B_5 on B_1 B_2 at B_4 such that its length is equal to that taken along A_4 - A_5. Extend the line (as dashed line) B_4– B_5 on either sides which will be parallel to BOB'(Fig. 6.1.1.). The extended B_4 - B_5 line stands for the "*H*" structure contour for the fault plane.

The two dashed lines A_4–A_5 and B_4– B_5 are named "*H*" structure contours because these stand for the projections of the occurrences of the sandstone and the fault at a depth of "*H*". These structure contours intersect each other at point O'.Join OO' and extend it to *C*. The line $OO'C$ gives the trend of the line of intersection of the sandstone and the fault plane (Fig. 6.1.1).

Step III Determination of Bearing and Plunge of the Line OOC

The direction of line $OO'C$ measured with respect to north line, gives the bearing of the line of intersection of the sandstone and the fault plane. It is found to be N 14 °E (Fig. 6.1.1). Point $O'$is "*H*" unit lower than point *O*, and this lowering is over the horizontal distance OO' Therefore draw a line OO'at a suitable place (Fig. 6.1.2) and then erect a perpendicular atO'

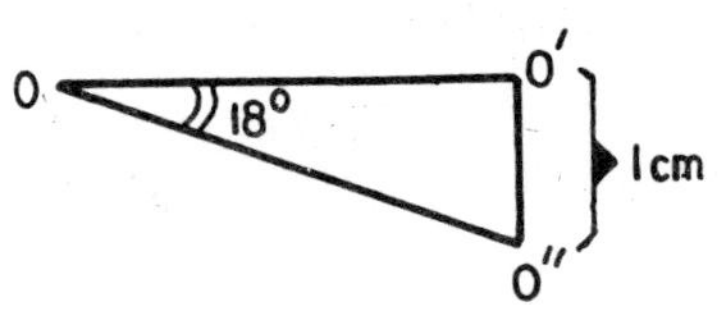

Fig. 6.1.2

Fig. 6.1.2. **CONSTRUCTION TO MEASURE PLUNGE OF LINE OOC (Problem 6.1)**

NOTE Draw $OO'$of same length as obtained in Fig. 6.1.1. Erect a perpendicular $O'O''$ at O'such that it is 1 cm. long. Join OO'' and measure angle $O'OO''$. It is found to be 18.

on OO'. The length of this perpendicular (OO' in Fig. 6.1.2) should be equal to "H" unit i.e., equal to length A_4–A_5 or B_4–B_5 (Fig. 6.1.1). Joint OO'' and measure angle $O'OO''$ which is a measure of the plunge of the line $OO'C$. This is called the plunge and its value is found to be 18°. (angle $O'OO''$ in Fig. 6.1.2).

6.1.2 (b) Data given as a Gradient

Problem 6.2 A sandstone dips 1 : 5 due N 20 °W, while fault dips 1 : 2 due S 40 °W. Find the bearing and the plunge of the line of intersection of these two dipping surfaces, by the geometric method.

Procedure

With the data given in the form of a gradient, the construction of "H" structure contour is rendered very easy, because it is now not necessary to construct the angle of dip with respect to the concerned direction of dip, as was required in the earlier case. The rest of the procedure is same as that described for the data based on inclination given in degrees.

Step I Drawing the Sketch of Intersection of the two Surfaces

Derive the direction of the strike from the given direction of dips namely N 20 °W and S 40 °W. The strike directions will

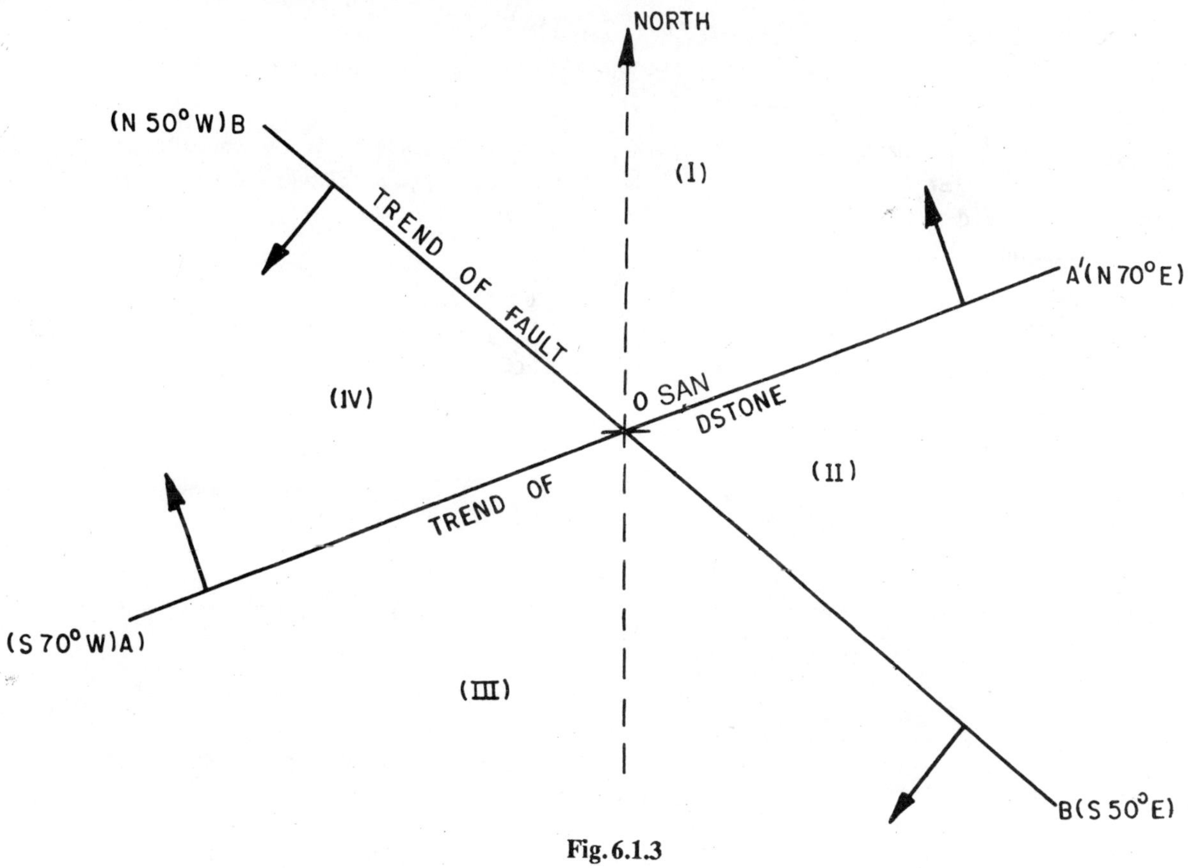

Fig. 6.1.3

Fig. 6.1.3. PLOTTING OF TRENDS, DIRECTIONS OF DIP OF SANDSTONE AND FAULT (Problem 6.2)

AOA' = trend of sandstone = N 70 °E - S 70 °W

BOB' = trend of fault = N 50 °W - S 50 °E

Note that both the surfaces dip toward the fourth sector and the second point of intersection is to be expected in that sector.

be N 70 °E – S 70 °E (sandstone) and N 50 °W – S 50 °E (Fault). Plot these directions as lines *AOA'* and *BOB'* and indicate the direction of dip (Fig. 6.1.3). Ascertain the sector in which the second point of intersection is to be expected. From Fig. 6.1.3. it is found that it is located in Sector IV.

Step II Construction of "*H*" Structure Contours

Erect a perpendicular A_1– A_2 on *AOA'* at *A*1, its length being equal to 5 units (because the sandstone dips 1 : 5).

Through A_2 draw a dashed line parallel to *AOA'*, this line through A_2 giving the "*H*" structuree contour of sandstone. Similarly erect a perpendicular B_1 B_2 on *BOB'* at B_1, its length being equal to 2 units (because the fault plane has a dip of 1 : 2). Through B_2 draw a dashed line parallel to *BOB'*, this line through *B*2 giving the "*H*" structure contour of fault plane. These two "H" structure contours intersect at point *O'*. Join *OO'* and extend it to *C*, which stands for the line of intersection of the sandstone and the fault plane. (Fig. 6.1.4)

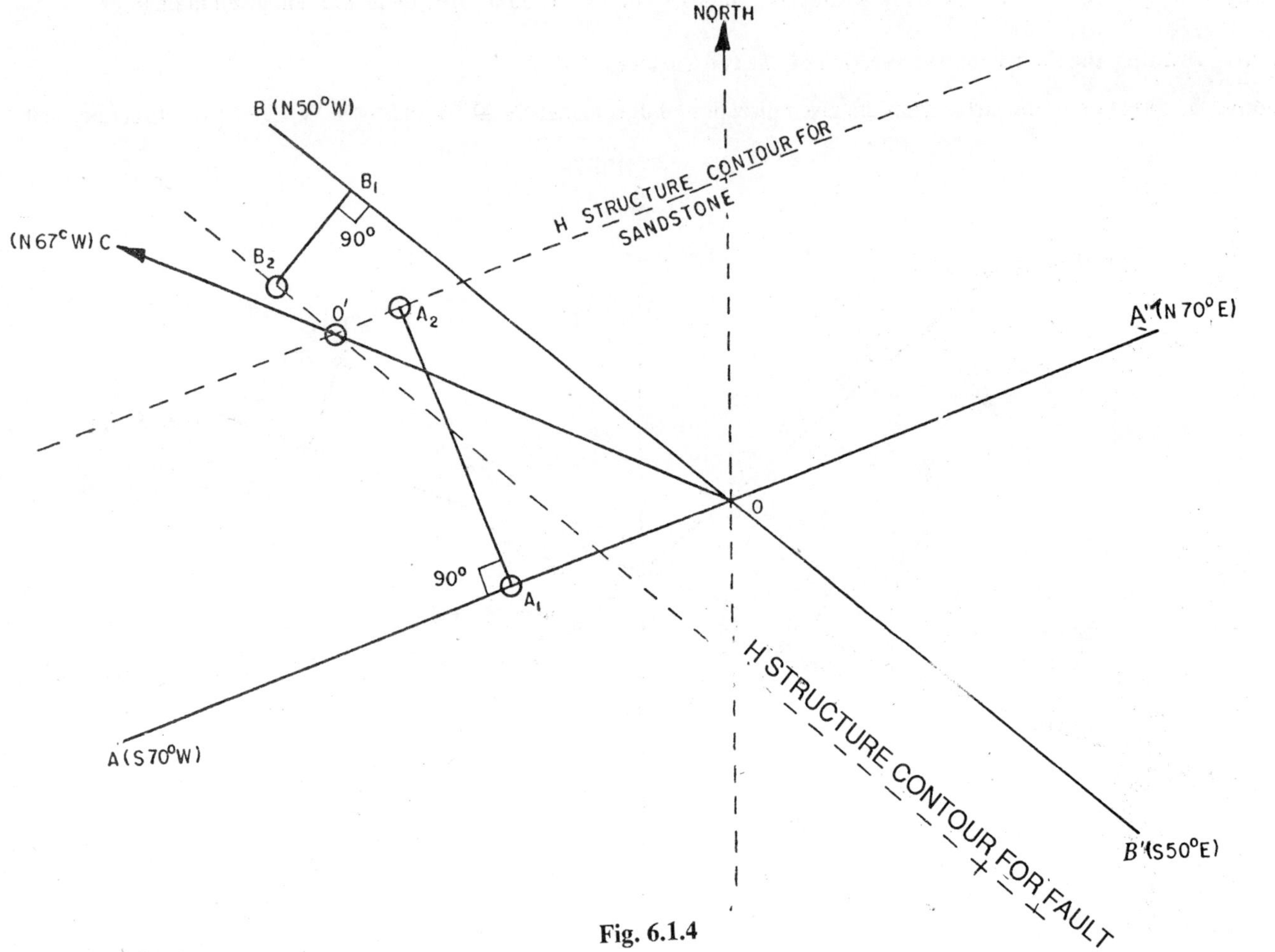

Fig. 6.1.4

Fig. 6.1.4 **CONSTRUCTION OF "H" STRUCTURE CONTOUR FOR SANDSTONE AND FAULT PLANE (Problem 6.2)**

A_1A_2 = perpendicular at A_1 on *AOA'*. A_1A_2 = 5 units

B_1B_2 = perpendicular at B_1 on *BOB'*; B_1B_2 = 2 units

Dashed line drawn through A_2 parallel to *AOA'* is the "*H*" structure contour for sandstone. Dashed line drawn through B_2 paralleel to *BOB'* is the "*H*" structure contour for the fault plane. These intersect at *O'*, which the intersection of the two surfaces at "*H*" depth.

$OO'C$ = line of intersection of sandstone and fault plane = N 67 °W

OO' = 7.3 units, which is the plunge of loine $OO'C$

Scale 1 cm = 1 unit

Step III Measureing the Bearing and the Plunge of the Line $OO'C$

$OO'C$ is the line of intersection of the two dipping surfaces. The bearing is found to be N 67 °W (Fig. 6.1.4). Its plunge is found to be 1 : 7.3 because the length of the line OO' is found to be 7.3 units.

6.1.3 Use of "*H*" Structure Contours in Fault Problems

In fault problems, it is often necessary to determine the rake of the net slip, as well as the down throw side. This is possible if two intersecting dipping bodies or horizons, like quartz veins, dykes, etc., are available along the fault plane. "H" structure contours are to be constructed for the dipping faulted bodies, for both the sides. From such a construction the trend of the net slip, rake of the net slip, down throw side etc., can be derived. The procedure for vertical fault planes is described in the following paragraphs.

Case I

Problem 6.3 A dolerite dyke trending N 30 °W – S 30 °E and having a dip of 40° due N 60 °E, and a basalt dyke trending N 50 °E – S 50 °W and having a dip of 60° due N 40 °W are affected by a vertical fault F_1– F_1 which has an E– W trend. Determine the bearing of the net slip, rake of net slip and the down throw side of the fault. The ground surface is flat.

Procedure

The positions of the faulted dykes along the fault is very important, as it controls the down throw side, bearing of the net slip and so on. Therefore for such problems a sketch of the positions of the fault, the faulted bodies etc., is required to be necessarily given. Fig. 6.1.5 shows the positions of the dykes of dolerite and basalt along the fault F_1– F_1. In such problems, the dipping bodies intersect only on one side of the fault. In Figure 6.1.5, the entire data has been given, and the same is treated further

Step I Construction of "*H*" Structure Contour The Fault $F_1 – F_1$

Erect a perpendicular $F'– F''$("H" units = 2 cm) on fault F_1–F_1. Draw a line F''–F''' parallel to the fault line through point F''. This yields the "*H*" sturcture contour of the fault. As the fault plane is vertical, the construction is simple. The "*H*" contour structure is to be constructed on that side of the fault where the bodies are actually intersecting each other. In the present case it is on the southern side of the fault plane (Fig. 6.1.6)

The Dolerite Dyke

This dyke is shown as A_1–A_3 on the northern and as A_2–A_4 on the southern sides of the fault, respectively (Fig. 6.1.5). Erect a perpendicular $A' – A''$ on A_2–A_4 and construct an angle $A''A'A''$ of value 40°, because dolerite dyke has a dip of 40°. This perpendicular must be drawn on the side of the direction of dip of the concerned body (in this case it is due N 60°E).. Erect a perpendicular A"--A"' at A" such that its length is "H" units (2 cms). Extend A"--A"' till it touches the fault F_1–F_1 at $A5$. This line gives the "*H*" structure contour of the dolerite (Fig. 6.1.6). Erect a perpendicular A_5–A_6 from A_5 on "*H*" structure conour of fault. Join A_2 and A_6 and extend it to point *C* (Fig. 6.1.6).

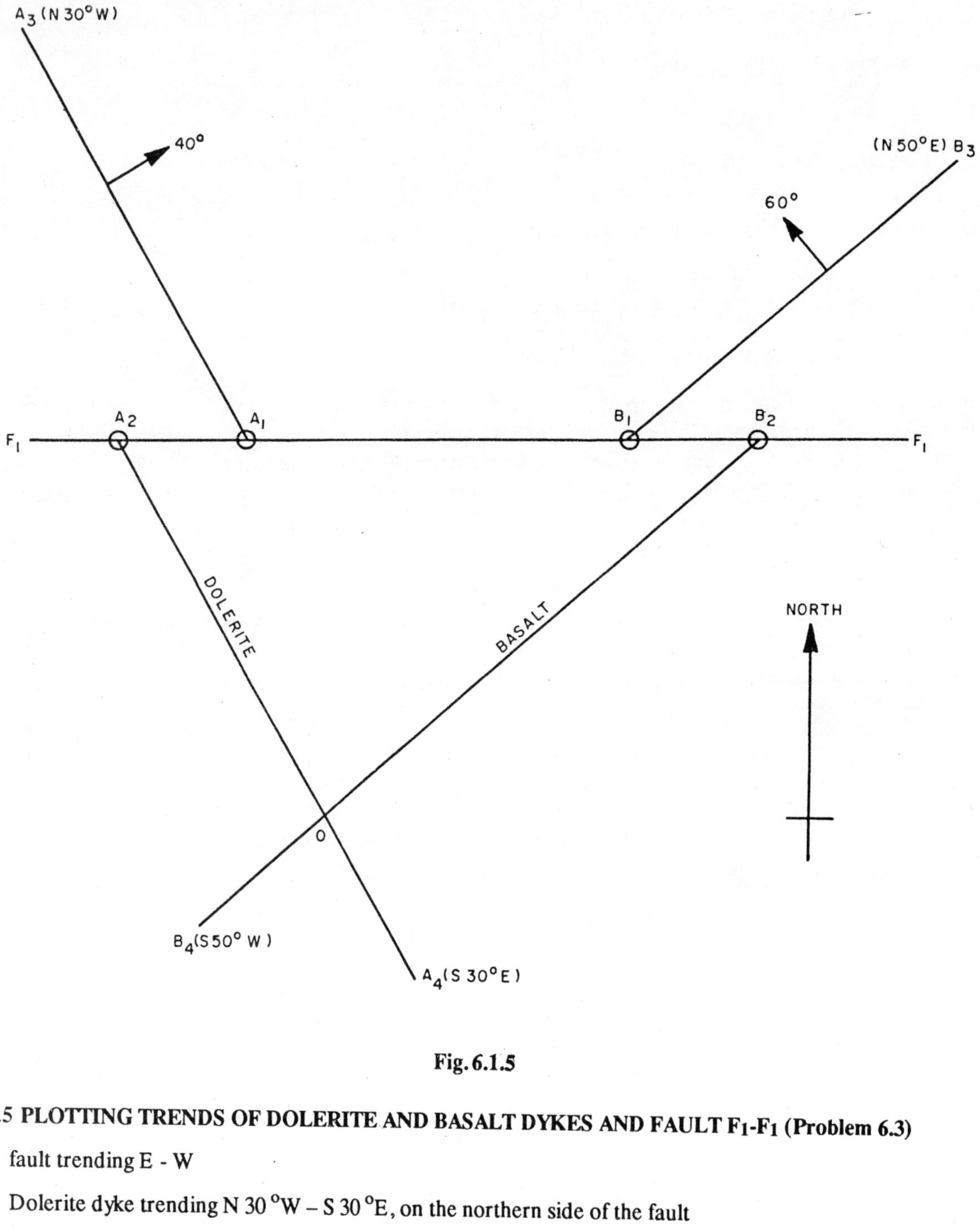

Fig. 6.1.5

Fig. 6.1.5 PLOTTING TRENDS OF DOLERITE AND BASALT DYKES AND FAULT F_1-F_1 (Problem 6.3)

F_1-F_1 = fault trending E - W

A_1A_3 = Dolerite dyke trending N 30 °W – S 30 °E, on the northern side of the fault

A_2A_4 = Dolerite dyke trending N 30 °W – S 30 °E, on the southern side of the fault

B_1B_3 = Basalt dyke trending N 50 °E – S 50 °W, on the northern side of the fault

B_2B_4 = Basalt dyke trending N 50 °E – S 50 °W, on the southern side of the fault

Note = The dykes intersect on the southern side of the fault.

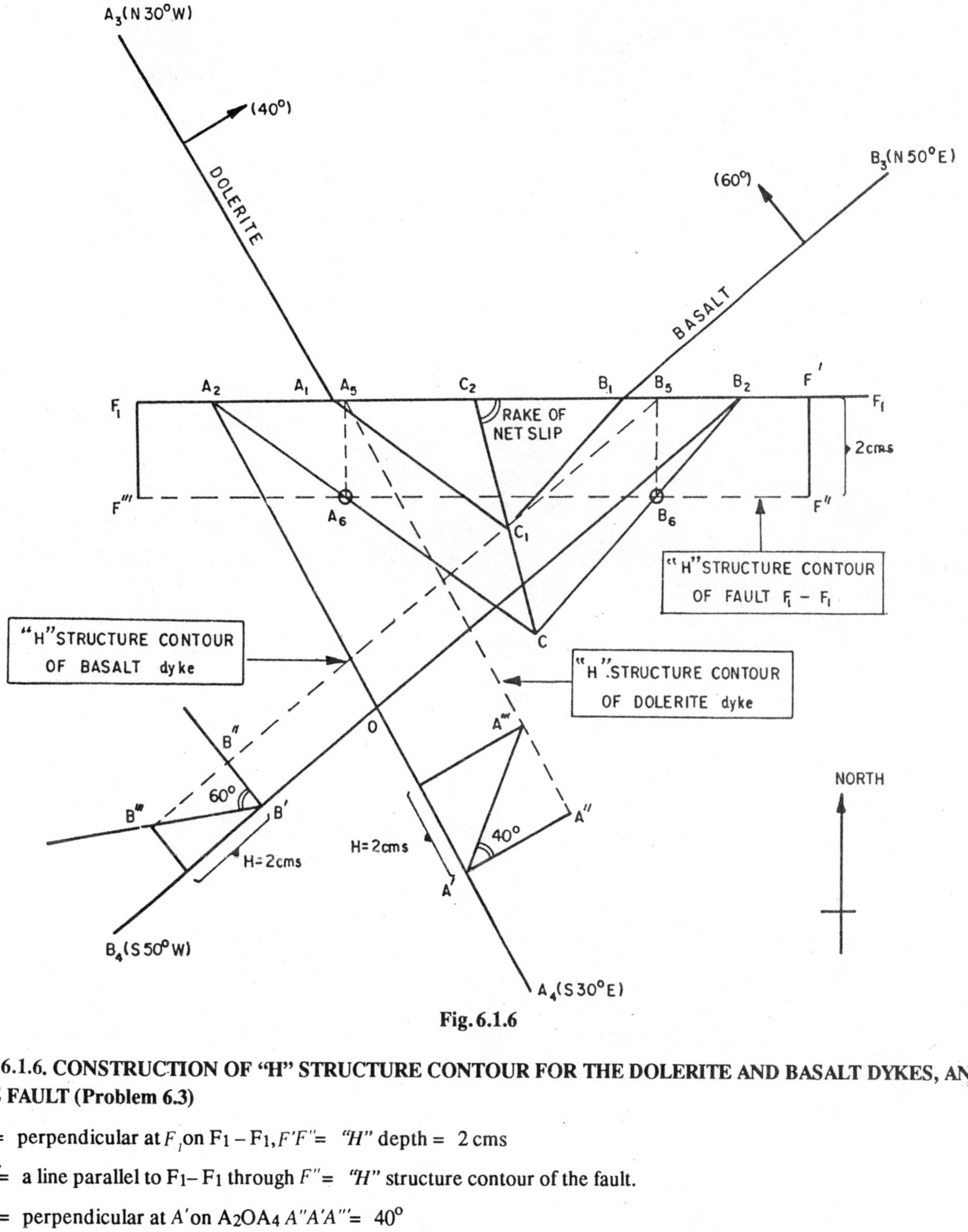

Fig. 6.1.6

Fig. 6.1.6. CONSTRUCTION OF "H" STRUCTURE CONTOUR FOR THE DOLERITE AND BASALT DYKES, AND THE FAULT (Problem 6.3)

$F'F''$ = perpendicular at F_1 on F1 – F1, $F'F''$ = "H" depth = 2 cms

$F''F'''$ = a line parallel to F1 – F1 through F'' = "H" structure contour of the fault.

$A''A'$ = perpendicular at A' on A2OA4 $A''A'A'''$ = 40°

$A''A'''$ = "H" depth = 2 cms

$A'A'''A_5$ = a line parallel to A2OA4 through A'' = H structure contour for dolerite dyke.

A_5A_6 = perpendicular at A_6 on $F''F'''$

A_2A_6C = trace of dolerite dyke occurring on the southern side of the fault, on the fault plane.

$B'B''$ = perpendicular at B' on B_2OB_4, $B''B'B'''$ = 60

B''' = "H" depth = 2 cms.

$B'''B''B_5$ = a line parallel to B_2OB_4, through B''' = "H" structure contour for basalt dyke

B_5B_6 = perpendicular at B_6 on $F''F'''$

B_2B_6C = trace of basalt dyke occurring on the southern side of the fault, on the fault plane.

A_1C_1 = a line parallel to A_2A_6C, through A_1. This is the trace of dolerite dyke occurring on the northern side of the fault, on the fault plane.

B_1C_1 = a line parallel to B_2B_6C, through B_1. This is the trace of basalt dyke occurring on the northern side of the fault, on the fault plane.

CC_1C_2 = trend of net slip of the fault

CC_2F_1 = rake of net slip = 75°

The Basalt Dyke

This dyke is shown as B_1–B_3 on the northern and as B_2–B_4 on the Southern sides of the fault, respectively (Fig 6.1.5). Erect a perpendicular B'–B'' on B_2– B_4 and construct an angle $B''B'B'''$ of value 60°, because basalt dyke has a dip of 60°. This perpendicular is drawn on the northwestern side of the dyke, as the dip is also on that side (Fig. 6.1.6). Erect a perpendicular B'''–B'' at B'' such that its length is equal to "H" units (2 cms). Extend B'''–B'' till it touches the fault F_1–F_1 at B_5 (Fig. 6.1.6). This line gives the "H" structure contour of basalt. Erect a perpendicular B_5–B_6 from B5 on the "H" structure contour of fault. Join B_2 and $B6$ and extend it to point C (Fig. 6.1.6).

Step II Determining the Trace of Dykes on the Fault

The lines A_2–A_6—C and B_2–B_6—C stand for the trace of dolerite and the basalt dykes on the fault plane F_1–F_1, respectively. However these form the southern side of the fault. Draw A_1–C_1 and B_1–C_1 parallel to A_2–A_6—C and B_2-B_6–C, respectively. These lines stand for the traces of dolerite and basalt dykes as are occurring on the northern side of the fault. The points C and C_1 yield the direction of net slip, and also the down throw side. It is held that the points C and C_1 were together prior to faulting.

Step III Determining the Trend of the Net Slip, Rake of Net Slip and down Throw Side

Join C and C_1 and extend it to point C_2 on the fault F_1–F_1. This line gives the trend of the net slip. This line is found to be running in a N 15 °W– S 15 °E direction. However the slip will be only in one direction and in this case it is from C_1 to C. Therefore the trend of bearing of net slip is S 15 °E (Fig. 6.1.6). The line C–C_1–C_2 meets at an angle to fault F_1–F_1, and the value is 85°. Therefore rake of net slip is also of 85°. Hence it is an "oblique slip fault".

Point C_1 stands for the intersection of the dykes on the northern side, while C stands for the intersection on the southern side. These two points were together prior to faulting. Point C_1 is at a shallower depth, while C is at a greater depth. Therefore southern side is at a lower level than the northern side. Therefore southern side is the down throw side.

IMPORTANT OBSERVATION

The outcrop lines of dolerite and basalt dykes on the northern side of the fault, are in between those on the southern side of the fault (Fig. 6.1.5). As a result, southern side is observed to be the down throw side.

Case II

Problem 6.4 In a region, dolerite and basalt dykes are obsrved to have been affected by a fault $F_1 - F_1$. The dolerite dyke trends N 30 °W – S 30 °E and dips 50° due N 60°E. The basalt dyke trends N 50 °E – S 50 °W and dips 60° due N 40 °W. F_1–F_1 is a vertical fault which is trending E – W. Determine the bearing of the net slip and the down throw side of the fault. The data is given in the form of a sketch (Fig. 6.1.7).

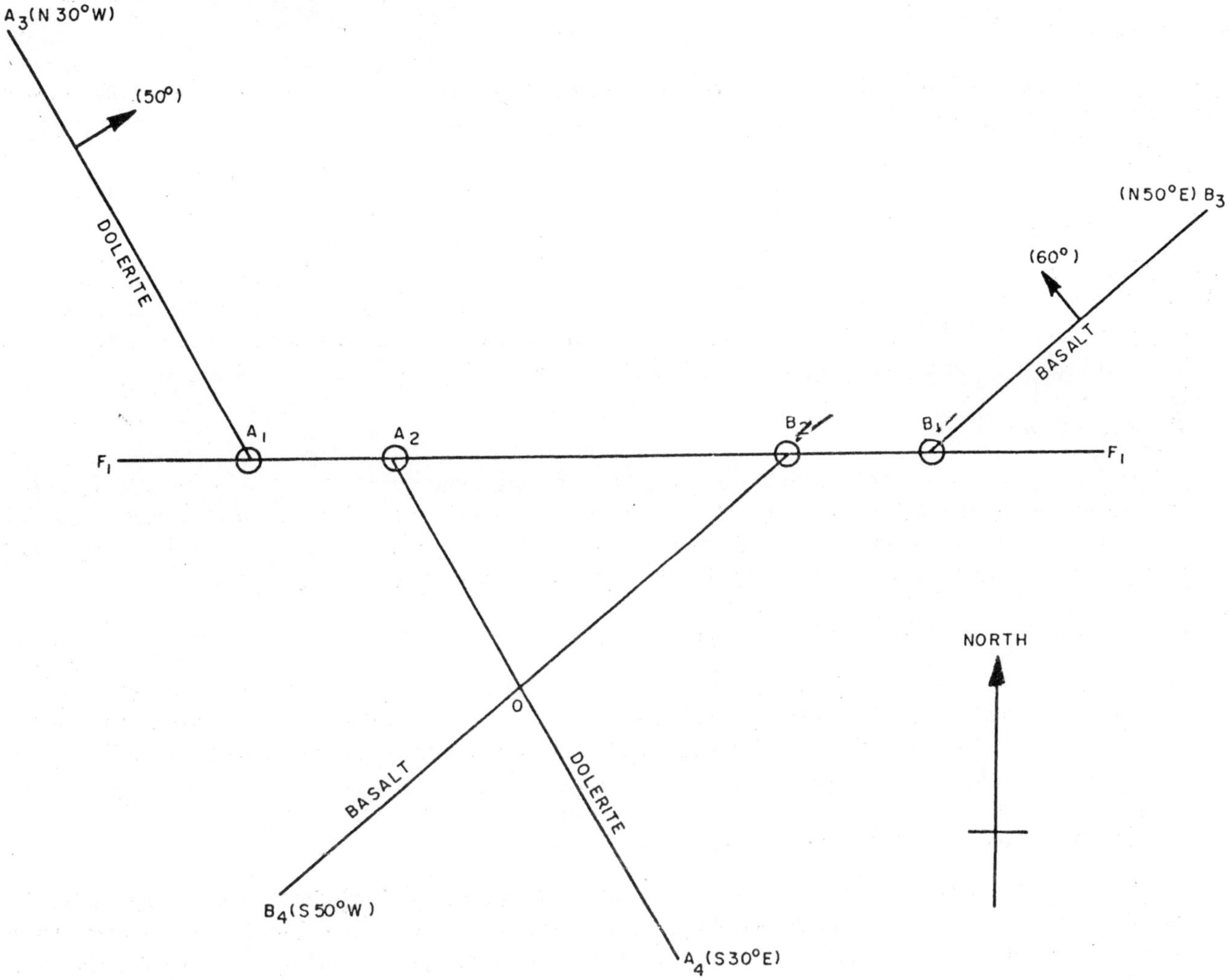

Fig. 6.1.7

Fig. 6.1.7. PLOTTING OF TRENDS OF DOLERITE AND BASALT DYKES, AND FAULT (Problem 6.4)

$F_1 - F_1$ = fault trending E-W

A_1A_3 = dolerite dyke trending N 30 °W – S 30 °E on the northern side of the fault.

A_2A_4 = dolerite dyke trending N 30 °W – S 30 °E on the southern side of the fault.

B_1B_3 = basalt dyke trending N 50 °E – S 50 °W on the northern side of the fault.

B_2B_4 = basalt dyke trending N 50 °E – S 50 °W on the southern side of the fault.

Note: The two dykes intersection the southern side of the fault.

Procedure

Step I Construction of "*H*" Structure Contour for Dykes and Fault

This is done only for dykes occurring on the southern side of the fault, because here the dykes are actually intersecting one another.

The Fault

The construction of "*H*" structure contour is very simple as the fault is vertical. Erect a perpendicular F_1–F'' at F_1 such that it is 2 cms. long and draw a line F''–F''' parallel to F_1–F_1 through F''. The line F'''–F'' gives the "*H* structure contour for the fault (Fig. 6.1.8).

The Dolerite Dyke

This dyke is shown as lines A_1–A_3 and A_2–A_4 in Figure 6.1.7. "*H*" structure contour is be drawn for A_2 - A_4. Erect a perpendicular A'''–A'' at A'' and construct an angle $A'A''A'''$ of 50° (Fig. 6.1.8). Erect a perpendicular A''- A'' at A'', such that it is "*H*" units long (2 cms). Extend the line A''–A'' upto A_5, which gives the "*H*" structure contour for dolerite dyke. Erect a perpendicular A_6–A_5 at A_6, which also should be 2 cms long. Join A_2 and A_6 and extend it to C (Fig. 6.1.8).

The Basalt Dyke

This is shown as lines B_1–B_3 and B_2–B_4 in Fig. 6.1.7. "*H*" structure contour is to be drawn for B_1– B_4 only. Erect a perpendicular B'- B "at *B* and construct angle $B''B''B'$ equal to 60°. Erect a perpendicular B''–B'' at B'', such that it is "*H*" units long (2cms). Extend the line B''–B'' to B_5, which gives the "*H*" structure contour for basalt dyke. Erect a perpendicular B_6–B_5 at B_6 which also should be 2 cms long. Join B_2 and B_6 and extend it to *C* (Fig. 6.1.8).

Step II Constructing the Trace of Dykes on the Fault Plane

The lines A_2– A_6—C and B_2–B_6—C stand respectively for the traces of dolerite and basalt dykes on the southern side of the fault F_1–F_1. Likewise A_1–C_1 (this is to be drawn parallel to A_2–A_6—C) and B_1–C_1 (this is to be drawn parallel B_2–B_6–C) stand respectively, for the traces of dolerite and basalt on the northern side of the fault F_1–F_1. It is argued that the points C and C_1 were together prior to faulting (Fig. 6.1.8).

Step III Determining the Trend of the Net Slip and down Throw Side

Join C_1 and *C*, and extend it to point C_2. This gives the trend of the net slip. This line is found to be running nearly in N–S direction (N 2°E– S 2°W). However the net slip will be only in one direction i.e., from *C* to C_1, north to south. The line C_1–C_2 is practically perpendicular to the trend of the fault F_1– F_1 (it is actually 88°). Therefore rake of the net slip is nearly 90° and hence it is a "dip slip fault". (Fig. 6.1.8).

Point C_1 is at a lower depth than *C* and these points correspond respectively to northern and southern parts of the fault F_1–F_1. Therefore the northern side of the fault is the down throw side.

IMPORTANT OBSERVATION

The outcrop lines of the dolerite and basalt dykes on the southern side are within those of the northern side and nothern side is the down side.

Thus from the positions of the outcrops, the down throw side or up throw sides can be inferred. This has been schematically shown in Figs. 6.1.9 and 6.2.0.

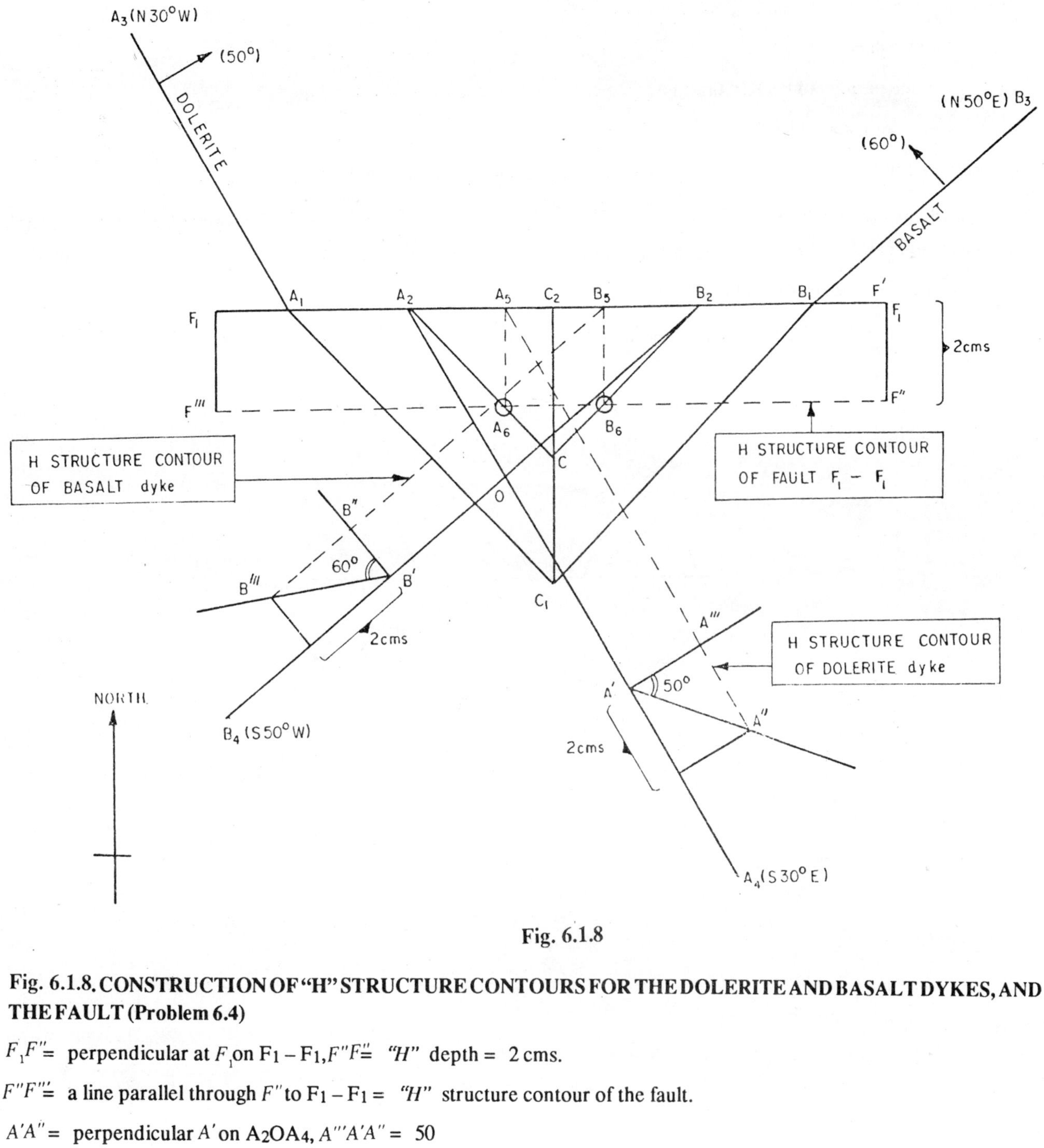

Fig. 6.1.8

Fig. 6.1.8. CONSTRUCTION OF "H" STRUCTURE CONTOURS FOR THE DOLERITE AND BASALT DYKES, AND THE FAULT (Problem 6.4)

F_1F'' = perpendicular at F_1 on F1 – F1, $F''F''$ = "H" depth = 2 cms.

$F''F'''$ = a line parallel through F'' to F1 – F1 = "H" structure contour of the fault.

$A'A''$ = perpendicular A' on A2OA4, $A'''A'A''$ = 50

$A''A'''$ = "H" depth = 2 cms.

$A''A'''$ = a line parallel to A2OA4, through A'' = "H" structure contour for dolerite dyke

A5A6 = perpendicular at A6 on $F''F'''$

A2A6C = trace of dolerite dyke occurring on the southern side of the fault, on fault plane.

$B''B'$ = perpendicular at B' on B_2OB_4, $B''B'B'''$ = 60°

$B'''B''$ = "H" depth = 2 cms.

$B'''B''B_5$ = a line parallel to B_2OB_4, through B'' = "H" structure contour for basalt dyke

B_5B_6 = perpendicular at B_6 on $F''F'''$

B_2B_6C = trace of basalt dyke occurring on the sourthern side of the fault, on fault plane.

A_1C_1 = a line parallel to A_2A_6C = trace of dolerite dyke occurring on the northern side of the fault, on the fault plane.

B_1C_1 = a line parallel to B_2B_6C = trace of the basalt dyke occurring on the northern side of the fault, on the fault plane.

C_1CC_2 = trace of net slip of the fault

$C_1C_2F_1$ = rake of the net slip = 90°

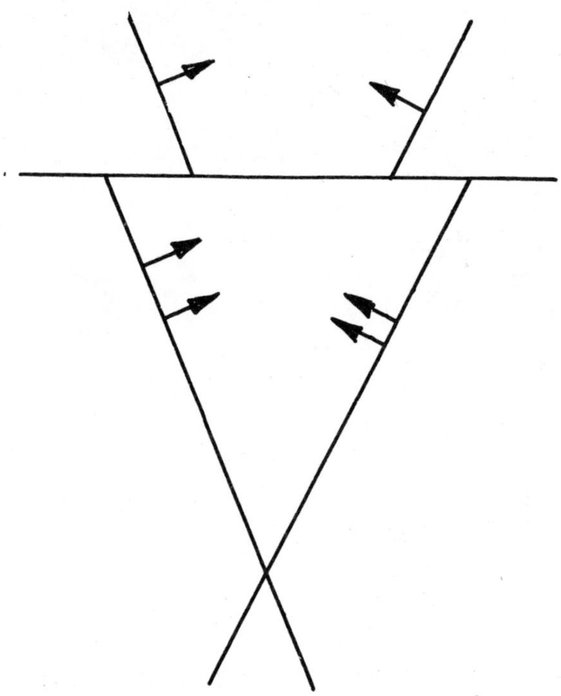

Fig. 6.1.9

Fig. 6.1.9 DISPOSITION OF OUTCROPS OF FAULTED DIPPING BODIES WHEN SOUTHERN SIDE IS THE DOWN THROW SIDE (Problem 6.3)

NOTE The outcrops of the bodies occurring on the northern side of the fault are within those occurring on the southern side. Therefore from such a disposition, the down throw side can be judged, but the amount of throw has to be determined by calculations.

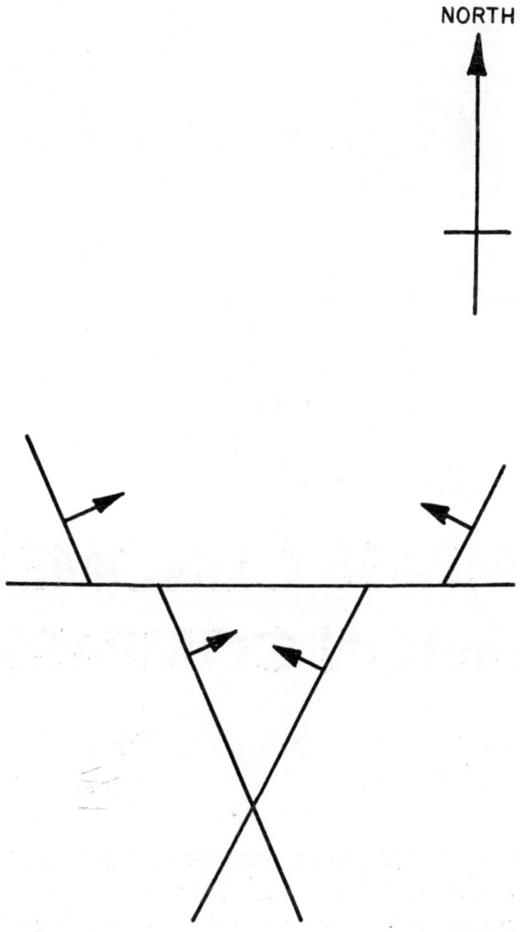

Fig. 6.2.0

Fig. 6.2.0 DISPOSITION OF OUTCROPS OF FAULTED DIPPING BODIES WHEN NORTHERN SIDE IS THE DOWN THROW SIDE (Problem 6.4)

NOTE The outcops of the bodies occurring on the sourthern is of the fault are within those occurring on the northern side. Therefore northern side is the down throw side. The amount of throw is required to be calculated, but from the distribution of the outcrops, the down throw has to be determined by calculations.

7

USE OF COMPASS CLINOMETER AND BRUNTON COMPASS

7.1 INTRODUCTION

The compass clinomter or Brunton compass is of paramount importance to geoscientists, since field data pertaining to recording the attitude of the rocks, trends of the faults, shears or any other linear elements, bearings and back bearings of objects etc., is possible only with these instruments. The desired direction of traverse in the field, say due to S 80 °W can be obtained only with the compass clinomter. It is therefore necessary that the geoscientist is familiar with the parts of the "compass" and the procedure to be adopted while recording data.

7.1.1 Parts of the Compass Clinomter/Brunton Compass

Essentially it consists of a magnetic needle placed on a sharp pivot such that the needle can rotate freely, when the pivot is made vertical, or the base of the compass is made truly horizontal. Such a needle moves over a graduated horizontal disc, the position of the needle at any place giving the direction of trend of a line, bearing or back bearing of objects sighted etc. Unlike a simple compass (like the trough or Mariners compass), there is a pendulum (in the case of compass clinometer) or an index (in the case of Brunton compass), with the help of which the inclination of any surface can be measured. The magnetic needle and the pendulum are housed in a non-magnetic case.

7.1.2 The Magnetic Needle

This is made to rest over a pivot, the latter being fixed at the centre of a graduated circular base. The southern end of the needle is made heavy by tieing some heavy non-magnetic material to it, because in the northern hemispehere of the earth, the north pole of the needle will be pulled down (this is called as magnetic declination). Such compass clinometer or the Brunton compass should be used only in the northern hemisphere. Likewise, for instruments to be used in the southern hemisphere, the northern end of the needle is made heavy, because now the south pole of the needle will be pulled down. By so doing, the magnetic declination is nullified. The north pole of the needle is either indicated by colour or in any other device, so that the readings of the bearings can be taken quickly. The north pole only is made prominent, because in most of the surveys, the bearing of a point is needed. Such a reading is obtained only with respect to the north pole of the needle.

7.1.3 The Graduated Disc/Base

The bearings, the back bearings, the trends etc., are taken with respect to the cardinal directions of North-South. All angular measurements are made with respect to the direction of the north. Therefore the position of the north line is foundamental in the instrument. The bearings or the directions can then be described as north, northeast, east, southeast, south, southwest, west and northwest etc. Also the total angle described starting from north and passing through east, south, west and back to north is of 360 degrees. Thus the bearings or the directions can be described as 20°, 90°, 140°, 180°, 250°, 300° and so on. The disc of the compass is therefore graduated in the following two ways:

(a) The quadrantal type

(b) The circular type

7.1.4 The Quadrantal Type of Graduations

The total angle of 360° is split into four quadrants namely, northeast, southeast, south west and northwest. At north and south the readings are "0°" (zero) while at east and west it is "90. This has been shown in Figure 7.1.0.

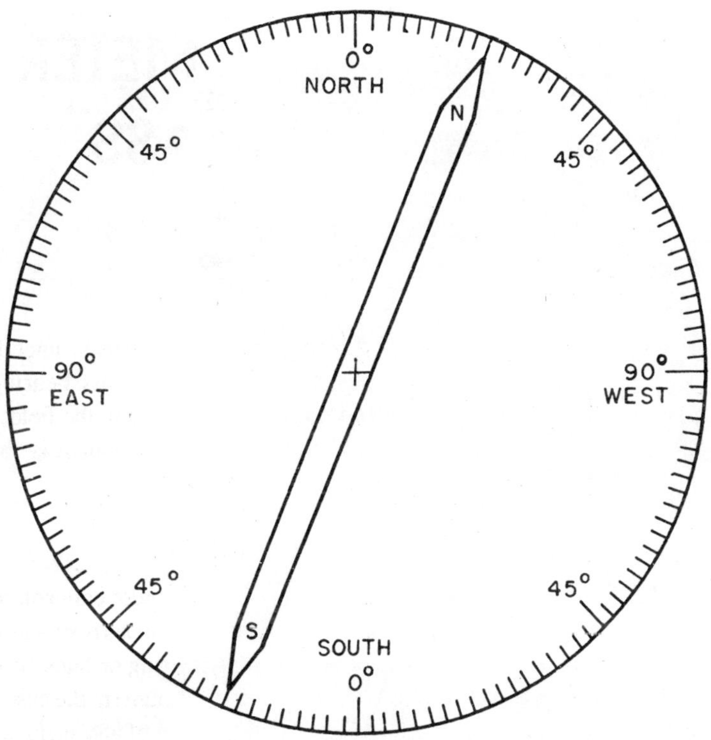

Fig. 7.1.0

Fig. 7.1.0 Graduations in quadrantal type. Note that west is marked on the right hand side, and east is marked on the left hand side of the N–S line

In each quadrant, further subdivisions are made at angles of 22° 30, 45° and 67°30, like *NNE* (22°30), E (45°), ENE (67°30).

The subdivisions in the other quadrants are *NNW, NW, WNW* (*NW* quadrant), *SSE, SE, ESE*, (*SE* quadrant), and *SSW, SW, WSW (SW* quadrant).

The quadrantal bearings are read in a specific way. Thus a reading N 67°W is read as 67° west of north, S 10°E is read as 10° east of south and so on. It will be thus clear that the readings are taken with respect to the cardinal directions of North or South.

7.1.5 The Circular Type of Graduations

The total angle of 360° is spilt into readings like 10°, 40°, 70°, 140°, 160°, 190°, 250° 280° 350° and so on. This has been shown in Figure 7.1.1.

From Fig. 7.1.1, it is seen that the readings are written in an anticlockwise manner, i.e., the readings increase in that direction upto 360°.

Study of Fig. 7.1.0 and 7.1.1, reveals that the correspondence between the two systems of graduations is as shown below.

Quadrantal bearing	Circular bearing
North	0° or 360°
N 15°W	345°
West	270°
S 45°W	225°
South	180°
S 45°E	135°
East	90°
N 45°E	45°

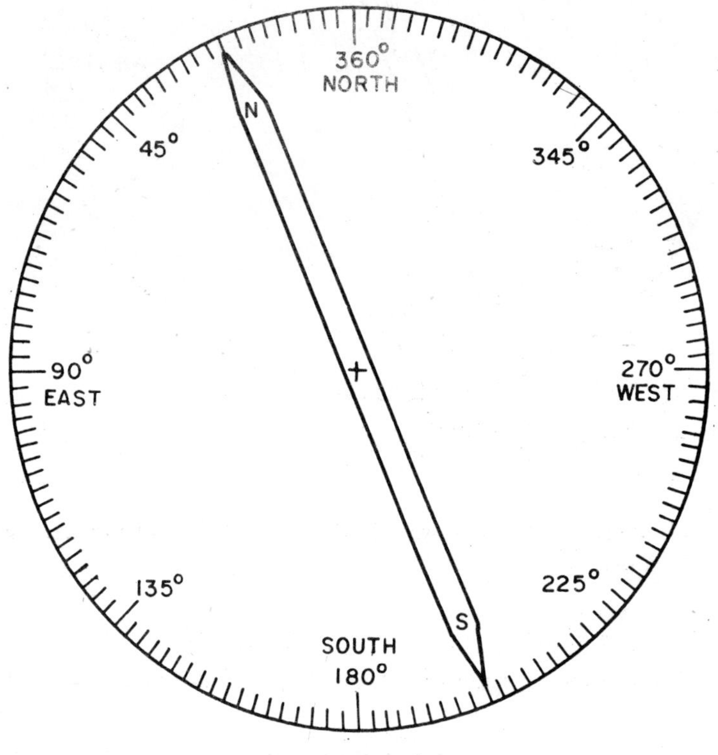

Figs. 7.1.1

Fig. 7.1.1 Graduations in circular-bearing type. Note that in this system too west is marked on the right hand side and east is marked on the left hand side of the N–S line.